HARVARD ECONOMIC STUDIES

Volume CXXII

The studies in this series are published by
the Department of Economics of Harvard University.
The Department does not assume responsibility for
the views expressed.

COMPETITION
IN THE
MIDWESTERN
COAL INDUSTRY

REED MOYER

Harvard University Press

CAMBRIDGE, MASSACHUSETTS

1964

79090

Library of Congress Catalog Card Number: 65–11194
Printed in the United States of America

FOR SUE

Preface

This study has grown out of the author's conviction that most of the existing economic analyses of the bituminous coal mining industry are out of date. The public's image of the coal industry reflects the chaos that overwhelmed the industry during the turbulent depression years of the 1930's. Scholars, depending on the existing literature for their knowledge of the industry, perpetuate what are now largely myths. Some fresh research obviously is sorely needed. This book seeks to fill a part of that need.

The post-World War II years have wrought great changes in the coal industry. A wave of crippling strikes in coal during the 1940's thrust the industry into the limelight. In the following years the industry slipped into a period of decline while gas and oil earned increasingly large shares of the energy market. But in recent years the industry has been gradually transformed and is offering gas and oil stiffer competition. Nowhere has the transformation been greater than in the midwest, the area on which this book focuses.

Unfortunately, time does not stand still to accommodate an author who tries to keep a book manuscript up to date. Merger activity continues, concentration ratios change, new technological developments are introduced. Except for the information in Table 32, which includes 1963 data, the events and data included in the book cover the period through 1962.

The list of people and agencies to whom I am indebted for help with various aspects of the research and writing of this book is long. I appreciate the cooperation of Mr. E. C. Hill of Tennessee Valley Authority in providing price information on T.V.A. coal bids. A number of coal executives and trade association officers gave valuable assistance, and to them I express my appreciation. I want, especially, to single out: Mr. L. E. Sawyer of the Indiana Coal Association, Dr. Myles Robinson and the late Carroll Hardy of the National Coal Association, Messrs. L. S. Weber, Harold V. Scott, and Julian E. Tobey of the Midwest Coal Producers Institute, Inc., Mr. Roy E. Dean of the Ayrshire Collieries, Mr.

S. F. Sherwood of the Stonefort Coal Mining Company, Mr. Jack Price of the Truax–Traer Division of Consolidation Coal Company, Mr. A. J. Christiansen, and the numerous other coal executives who generously responded to the confidential survey referred to in Chapter VI.

A special word of appreciation is due Professor C. Lawrence Christenson of Indiana University, truly "a gentleman and a scholar." Professor William W. Haynes of the University of Kentucky and representatives of the Kentucky Department of Revenue were helpful in providing confidential income and balance sheet data submitted by Kentucky coal operators, and to them go my thanks. Complying with the Department's request, data were grouped to prevent disclosure of confidential information pertaining to individual operators. I acknowledge also a debt to Miss Doris Reed and the staff of the Lilly Library at Indiana University for providing access to the Indiana Coal Association's coal distribution records. Mr. Martin Frederixon helped with some of the details involved in turning the manuscript into a book. Miss Patricia Moyer assisted in preparing the Index.

Three men deserve special mention for their substantial intellectual contributions to this book. I am particularly indebted to Professor Joe S. Bain for his guidance and stimulation. Through his counsel he improved the manuscript at an early stage, when he served as chairman of the author's doctoral dissertation committee. Professor Richard B. Heflebower also made many helpful suggestions which have been incorporated into the book. In addition, I owe Professor Richard E. Caves a debt of gratitude for his perceptive comments at several stages of the book's history. He is largely responsible for improvements in the book's organizational structure.

I am grateful to the University of California for a small travel grant and to Professor Thomas A. Staudt of Michigan State University for securing research funds.

I reserve for last my greatest debt of gratitude to my wife, to whom I dedicate this book.

East Lansing, Michigan   R.M.
July, 1964

Contents

Tables

xii   TABLES

Figures

CHAPTER I

Introduction

Most previous studies of the coal industry fall into one of several categories. They are either descriptive, general treatments of the industry as a whole, technical books covering combustion, mining technology and geology, analyses of the coal resource base, or studies of frequent government intervention in the industry's affairs. Most of these published works on coal are characterized by two features: (1) They are outdated. (2) They are devoid of economic analysis. Several important exceptions qualify these generalizations. Within the past decade a number of significant resource base studies by the Paley Commission and by Sam Schurr, Bruce Netschert, et al. have contributed to our understanding of the role of coal in a high-energy economy.[1] James Henderson's recent and interesting linear programming model of the industry, with which he derives "efficient" output solutions, represents a major advance in economic analysis.[2] C. L. Christenson's thorough study charts the "redevelopment" which has reshaped the industry in the post-World War II period.[3]

But the prevailing view of the industry is an outmoded one. In part this view stems from the obsoleteness of the existing works of the industry. Their obsolete character is partly determined by the industry's shifting fortunes. Once a dominant force in American industry, coal has declined in relative im-

---

[1] *Resources for Freedom,* President's Material Policy Commission, June, 1952; Sam H. Schurr and Bruce Netschert, *Energy in the American Economy, 1850–1975* (Baltimore: The Johns Hopkins Press, 1960).

[2] James M. Henderson, *The Efficiency of the Coal Industry* (Cambridge: Harvard Univ. Press, 1958).

[3] C. L. Christenson, *Economic Redevelopment in Bituminous Coal* (Cambridge: Harvard Univ. Press, 1962).

portance as a fuel supplier in recent years. Attention has shifted to newer, more dynamic industries.

The paucity of rigorous, economic analysis is harder to explain. It is due in part to the unavailability of refined, continuous, economic data applicable to the industry. Additionally, it may stem from the differences that characterize different segments of the industry. This diversity among the industry's separate parts makes economic analysis somewhat difficult. Generalizations that accurately picture one coal district may be invalid for another.

The coal industry is composed of over 5,000 firms operating 8,300 mines in twenty-nine producing states. Production and marketing conditions vary significantly from one producing district to another. For example, the problems facing Pennsylvania producers, whose production has declined 61 percent in a generation, cannot be compared with the situation in western Kentucky, where output is increasing. Labor costs differ from district to district, depending on resource conditions and the degree of unionization. Some producing areas, relying heavily on sales to domestic consumers, have sales patterns with widely fluctuating seasonal variations; other districts reverse the normal seasonal pattern by shipping heavily into inventory during the summer season.

The structure of firms in different segments of the industry varies also, with profound effects on economic behavior. Some districts are composed of large numbers of small firms battling for survival in the face of declining markets; elsewhere, a constantly shrinking group of companies divides an increasing market. The structure of ownership differs in another important respect. Some areas depend heavily on sales in the commercial market, while others rely on captive markets. The lion's share of Alabama's twelve-million-ton production, for example, is captive to steel and electric utility plants in that area and never enters into the competitive market place. At the other extreme, virtually none of the production from Illinois, Indiana, and western Kentucky is captively controlled.

The problems of analysis raised by the difficulties outlined above were fully recognized in the *Report of the Committee on Prices in the Bituminous Coal Industry* published by the National

Bureau of Economic Research.[4] Its conclusions, reached in 1938, apply equally well today. The report states in part: "The preceding analysis suggests the complexity of the problems with which a program of research (in coal) must deal. In addition to the natural conditions, most of which vary from field to field and materially influence the cost of production, there are economic forces, many of which are more or less common to all coal fields but (which) vary markedly in duration, degree and intensity." [5] Recognizing that "certain areas have improved their position relatively, and during many years, absolutely," the report suggests that "research, to be effective, should concern itself at the outset primarily with major producing and marketing regions . . . rather than with the industry as a whole." [6]

The present study follows the National Bureau's recommendation. It is concerned with the coal-producing regions of Illinois, Indiana, and western Kentucky. In much of the analysis, these districts will compose an "industry." Whenever the problems involving coal have general application, however, the analysis will broaden to cover the national coal scene. This use of an elastic "industry" concept is perfectly justifiable. An industry can vary in size depending upon the variables under review. Demand elasticities are just as easily studied on a national basis as on an individual district basis; on the other hand, the numbers and concentration of firms in different regions differ sufficiently to define the industry on a narrower basis.

What is the value in putting an industry under the microscope? In this connection, Joe Bain sums up three contributions that studies of this type have made:

(1) They explore in considerable detail the fundamental institutional, technological and geographic conditions which lie behind "demand and supply" and affect enterprise behavior.

(2) A related contribution . . . is to reveal the complexity of competitive behavior and its determinants in actual situations, and thus to sug-

---

[4] New York: National Bureau of Economic Research, 1938.
[5] *Ibid.*, p. 27.
[6] *Ibid.*

gest reasons for wide divergences in observed behavior of industries, which would seem to fall within a single category in a priori theory.

(3) A third contribution . . . is of course in finding and measuring the significant results of competitive behavior in various industries.[7]

The present work joins the ranks of previous studies which have covered a wide range of industry classifications. It will permit a detailed study of the behavior of a large segment of an important American industry. And hopefully, the results of this study together with the results of other, similar studies will provide useful generalizations about the functioning of our economy.

Methodology

This study deviates from the approach of Edward H. Chamberlin which relates price-output decisions to the shapes and positions of demand-and-cost curves, empirically hard to define. Furthermore, our concept of a market and market structure differs from the traditional, theoretical definition. A. C. Pigou defines a market as a "common nodal point" where a product, units of which are perfectly substitutable for each other, are bought and sold.[8] Joan Robinson views the market as a geographic area or product area bounded by "a gap in the chain of substitutes." [9] Going beyond these narrow conceptions, we conceive a seller's market structure to include "all those considerations which he takes into account in determining his business policies and practices." [10] For effective analysis of an industry's output and price decisions, we need to broaden market structure to encompass more than simply the size and numbers of firms in the industry and the degree to which product differentiation influences buyer behavior. Implementing this view, we want to classify market structure on the seller's side by "grouping together those firms . . . which operate under

[7] Joe S. Bain, "Price and Production Policies," *A Survey of Contemporary Economics,* ed. Howard S. Ellis (Philadelphia: The Blakiston Co., 1949), II, 146–147.
[8] Referred to in Edward S. Mason, "Price and Production Policies of Large Scale Enterprise," *American Economic Review,* Supplement, XXIX (March 1939), 68.
[9] *Ibid.,* p. 69.
[10] *Ibid.*

the same or similar objective conditions." [11] Included in these conditions are the cost and production characteristics of major seller groups' operations, the economic characteristics of the product, ease of entry into the industry, in addition to the geographic delineation of the market's production and consumption areas, and numbers and relative sizes of buyers and sellers which the industry's firms must take account of when making their price and output decisions.

Within this broader framework, enlarging the environmental influences under study, we attempt to link structure, conduct, and performance causally, the central goal being the illumination of behavioral characteristics. How, we ask, do the firms' activities affect the degree of resource employment? How efficiently are these resources utilized? What are the patterns of income distribution?

Since the market structure, conduct, and performance framework supports the study, these concepts need defining. Market structure refers to "the organizational characteristics of a market," with emphasis on "those characteristics which determine the relation of sellers in the market to each other, of buyers in the market to each other, of the sellers to the buyers, and of sellers established in the market to other actual or potential suppliers of goods including potential new firms which might enter the market." [12] That is, it includes "those characteristics of the organization of a market which seem to influence strategically the nature of competition and pricing within the market." [13]

With this definition in mind, the investigation to follow studies these structural characteristics of the midwestern coal industry:

(1) Shapes of the market areas facing the midwest and reasons for changes in their dimensions.
(2) The relative isolation of midwestern coal markets and the impact of this phenomenon on price and behavioral characteristics.

[11] Ibid.
[12] Joe S. Bain, Industrial Organization (New York: John Wiley and Sons, Inc., 1959), p. 7.
[13] Ibid.

(3) The size distribution of buyers in various market segments.

(4) Demand elasticities together with the influence they exert on producers' price-output decisions.

(5) The degree of seller concentration and the size distribution of sellers in the midwestern industry.

(6) The ability of firms in the industry to differentiate their products physically and through the use of advertising or other sales promotion devices.

(7) Conditions of entry and exit in the industry with particular attention given to the structural influence of coal reserves ownership.

(8) The cost and production characteristics of the industry's operations.

The term "market conduct" alludes to the patterns of behavior which enterprises follow in adapting or adjusting to the markets in which they sell (or buy).[14] This work looks at conduct from several angles, studying:

(1) The extent, if any, of attempts to coordinate prices multilaterally by explicitly collusive means or, indirectly, through "conscious parallelism."

(2) The use of medium- and long-term contracts removing coal from the continuing pressures of the market.

(3) The use of price discrimination in various markets.

The concept of "market performance" refers to "the composite of end results in the dimensions of price, output, production cost, selling cost, product design, and so forth, which enterprises arrive at in any market as the consequence of pursuing whatever lines of conduct they espouse."[15] Here, the study deals with:

(1) Earnings' rates in the midwest, comparing them whenever possible with profits in the rest of the industry.

(2) The degree of productive efficiency existing in the midwest, as determined by scale, utilization, etc.

(3) The performance of the midwest with respect to resource conservation.

[14] Bain, p. 9.
[15] *Ibid.*, p. 11.

(4) The nature of the distribution of income among the factors of production.

(5) The effectiveness of resource allocation in the industry.

But the study does more than review end results; in addition, it looks at the various dimensions of performance in the light of their determinants. It studies the relation of the structure of seller concentration to earnings' rates, analyzing the correlations, if any, between mine size and profit levels. Similarly, it determines the influence of seller concentration on efficiency. Finally, it examines the effect of mining techniques on resource conservation.

The study carves out the midwestern coal-producing region from the rest of the industry, examining it within the framework outlined above. We justify this geographic excision by indicating the extent to which the midwestern producing regions and their markets are relatively isolated from competing coal fields. Also, an attempt is made to demonstrate the influence on behavior of this relative isolation. Furthermore, the hypothesis that the structural characteristics of the midwestern coal industry are causally linked to its conduct and performance is tested. Finally, the study reviews the effectiveness of competition within this segment of the industry, learning the extent, if any, to which it has broken away from the destructively competitive pattern which long has characterized the bituminous coal industry in the United States.

Data Limitations

It becomes quickly apparent to anyone studying economic questions relating to coal that there is a paucity of good current data in several important areas. Thanks to the Bureau of Mines and the states' Departments of Mines and Minerals, there are generally satisfactory production and employment statistics. But in the important areas of prices, profits, and costs, satisfactory current data are sadly lacking. During previous attempts to regulate the industry's affairs — notably during the National Recovery Act and Guffey Act periods — data collection improved considerably. Unfortunately for those interested in coal research, these government experiments were short lived and too far removed in time from most of today's problems to provide much currently useful material.

The authors of the *Report of the Committee on Prices in the Bituminous Coal Industry* acknowledged the problems involved in collecting coal price data. Their experiences indicated "that coal was one of the most difficult commodities for which to get uniform and reliable price quotations." [16] Unlike other extractive industries there are no posted prices for coal which lend themselves readily to analysis. Coal deposits are sufficiently heterogeneous to provide a wide range of prices. The problem is compounded by the existence of freight differentials among the various districts which must be reckoned with in setting mine prices. In addition, the industry practice of product differentiation through the proliferation of sizes produced has increased still more the number of prices prevailing in the market. Add to these considerations the market imperfections created by the inadequate knowledge on the part of some buyers and the unequal bargaining power of sellers and buyers in different situations, and the result is an incredibly complex structure of prices. Coals are amenable to classification on the basis of quality, size, seam, etc.; however, no agency now publishes price data on the detailed basis that such a classification scheme would dictate.[17] This kind of price information has not been publicly available since the Guffey Act, which regulated coal prices, expired during World War II. The scanty published coal price data available today are simply averages of the thousands of individual prices. Where we use these prices, the reader should recognize their limitations.

Satisfactory cost and profit data are equally hard to obtain. The industry for the most part is composed of small, family-owned firms, bound by tradition to a policy of secrecy. Freed from the disclosure rules that apply to firms with listed securities, the smaller, unlisted companies jealously guard their financial data. Nationally, only 29.7 percent of 1960 production was mined by companies which publish annual earnings' statements. In the midwest, the situation was improved but far from satisfactory:

[16] *Report on Prices in the Bituminous Coal Industry,* National Bureau of Economic Research, 1938, p. 12.
[17] Several regional trade associations publish detailed price information, but it is generally available only to their members on a confidential basis.

58.6 percent of the Illinois, Indiana, and western Kentucky 1960 production came from companies with listed securities.

Fortunately, however, a large amount of hitherto unavailable data has been secured and is here published for the first time. In some cases the data refer only to one or another of the three coal districts covered by this study; nevertheless, they are useful in examining issues common to the entire region. Confidential financial data for individual western Kentucky mines have been released by the Kentucky Department of Revenue, furnishing useful balance sheet and earnings' figures. Additional confidential records giving individual mine prices as well as subdistrict average prices in Indiana and Illinois have been made available by individual operators and trade associations. Detailed production cost records have been secured through the generosity of a major midwestern producer. Additional data came from numerous coal company executives and trade association officials interviewed in the course of the study, as well as from replies to a questionnaire mailed to midwestern producers.

As in all studies of this type, the investigator is faced with raw material limitations as frustrating as they are inevitable. A choice must be made either to proceed on the basis of limited data or, denied perfection, to terminate the work. In this instance we followed the first alternative. We hope that publishing the material that is available and providing fresh insights into the operation of this industry outweigh whatever disadvantages arise from some data limitations.

CHAPTER II

Midwestern Coal Markets

As indicated in the introductory chapter, this study assumes that a meaningful analysis of competitive relations in the coal industry requires segmenting the national coal market into its various submarkets. This book focuses attention on the midwestern coal industry and on the market area it predominantly serves. The present chapter delineates both the production and the consumption areas on which we fix our sights for most of the remainder of the book. Additionally, it studies the geologic and geographic factors that give the midwest its unique character.

Turning to the market side, it analyzes the importance of the coal freight-rate structure in sheltering midwestern coal markets from the breezes of competition from other coal-producing districts.

Location and Extent of Coal Reserves

Figure 1 locates the coal resources of the United States. Geographically, the reserves are divided into six major fields: Eastern, Interior, Gulf, Northern Great Plain, Rocky Mountain, and Pacific Coast. As a general rule, coal rank decreases progressively as one moves from east to west, the highest-quality coals being located in the Eastern province, the low-rank lignites in the west, and the intermediate low-rank bituminous and subbituminous in the Interior region.

The Eastern region follows the Appalachian Mountain chain from Pennsylvania to Alabama, running through eastern Ohio, eastern Kentucky, Virginia, West Virginia, and Tennessee. Coal quality varies from one section of the region to another, but the general level of quality is high. Here most of the country's cok-

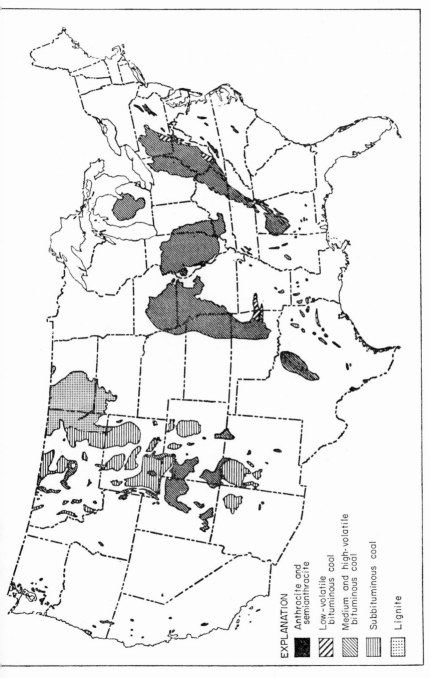

EXPLANATION

■ Anthracite and semianthracite

▨ Low-volatile bituminous coal

▧ Medium and high-volatile bituminous coal

▥ Subbituminous coal

▒ Lignite

FIGURE 1. Location of United States' coal resources

ing-coal reserves are located. Within the region, quality improves as one moves eastward, a condition resulting from the different pressures created when the mountains were formed.

The Interior province is made up of two major subdivisions: the Western Interior and the Eastern Interior fields. The Western field embraces deposits in Iowa, Missouri, Kansas, Oklahoma, and Arkansas. Quality, on the whole, is lower than that found in the Eastern Interior region, which is the area with which this study is primarily interested.

The Eastern Interior province blankets 67 percent of Illinois and much of southwestern Indiana and western Kentucky.[1] Here again quality differs from one part of the region to another, but the average lies between the inferior coal west of it and the high-rank Appalachian coal to the east. Generally, coal quality within the province improves toward the south, the best coals being located in the southern Illinois and western Kentucky fields. The Eastern Interior region holds a commanding advantage over most of the Eastern coal fields with respect to coal seam thickness and mining conditions in general. The significance of this important resource advantage will become apparent when later we discuss barriers to entry into the midwestern industry.[2]

Limited by low-rank reserves and separated by vast distances from the country's major industrial center, the remaining regions produce only 3 percent of the nation's coal. The Gulf coals, confined mostly to Texas, and coals of the Northern Great Plains area in the Dakotas, Montana, and Wyoming are mostly lignitic and subbituminous grades. The Rocky Mountain reserves vary more in quality, ranging from subbituminous to fairly high-quality, bituminous coking coal in Utah, supplying the California steel industry. Pacific Coast coal reserves are centered principally in Washington and play an insignificant part in meeting the nation's coal requirements. Virtually none of the western coal production directly competes with midwestern coal.

[1] The apparently large reserves in Michigan indicated on Figure 1 are illusory due to the irregular nature of the coal beds located there. Michigan coal production is negligible and never has been an important factor in the American coal industry.

[2] The terms "Eastern Interior" and "midwest" will be used interchangeably hereafter to refer to the coal-producing region of Illinois, Indiana, and western Kentucky.

Figure 2 shows in more detail the location of the Eastern Interior field. It constitutes a single, large, coal region, geologically united but politically divided by state boundaries. The field is shaped like an elongated saucer with the coal measures dipping gently from the perimeter toward the center. Mining is carried out for the most part toward the edges of the region where the seams are shallowest. But there are other important forces in-

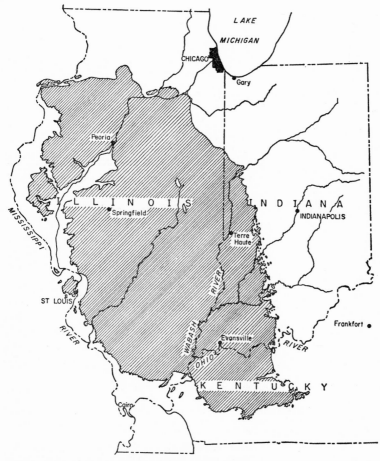

FIGURE 2. Location of Eastern Interior coal field

fluencing the location of mining activity in the midwest. We turn our attention now to the most pervasive of these forces, the market for midwestern coal.

Theory of Markets

The theoretical concept of a market is not complex. It can be defined as a "region in which buyers and sellers are in such free intercourse with one another that the prices of the same goods tend to equality easily and quickly." [3] In its simplest terms, the "region" is a single geographical point or, as in Harold Hotelling's conception, a line. Broadened, the concept of a "region" expands to a two-dimensional market area where producers sell to buyers located therein at prices influenced by the pressures of competitive sellers.

A curve that is the locus of points at which delivered costs from two producing points are equal defines the boundary between the market areas of two competing producers. The curve is the locus of points at which delivered costs from the two producing areas are equal. If the production costs for two producers are unequal, the higher-cost producer must be located closer to the boundary line to permit his lower transportation costs to offset his higher production costs. In this formulation, transportation rates are assumed to be proportional to distance, and average production costs are invariant with respect to the quantity produced. Products are homogeneous as well. Since the difference in transportation charges from any two producing locations (foci) to any point on the boundary line must equal the constant difference in production costs from the two producers, the differences in the transportation costs will likewise remain constant. The boundary line between market areas, therefore, is always a hyperbola.[4]

Several factors can contribute to the expansion of one-dimensional boundary lines into two-dimensional zones of indifference. Instead of there being a sharp division of market areas delimited

[3] E. A. G. Robinson, *The Structure of Competitive Industry* (London: Nisbet & Co., Ltd., 1935), p. 7.

[4] See Frank A. Fetter, "The Economic Laws of Market Areas," *Quarterly Journal of Economics*, XXXVIII (May 1924), 525. Exception: The boundary can be a straight line perpendicular to a line running between the two points *if* the f.o.b. prices are identical; otherwise they are hyperbolic curves.

by hyperbolic boundaries, the lines of demarcation can become blurred. These overlapping zones result from three causes. First, freight rates, instead of being directly proportional to distance may increase with distance in steplike progressions. On each of these rate plateaus, competitive overlapping can occur. Beyond the initial point on the plateau, the carrier absorbs transportation costs. The two other causes of overlapping zones result from buyers and sellers absorbing costs. Buyers can create zones of overlap from disparities in their preference patterns that permit each to assign different values to products that have been objectively classified as homogeneous. Sellers' absorptions stem from price cuts that allow them to invade markets beyond their "natural" market area. Since demand in fringe coal markets is highly elastic, it is to the sellers' advantage in these markets to shade price in the hope that their sales will expand.

In some cases the burden of price competition can be shifted from the seller to the transportation agency when producers cut delivered costs by instigating freight-rate reductions, eschewing the traditional procedure of reducing selling prices. Interestingly, some of the most bitter—and certainly the most continuous—freight-rate struggles between competing coal districts have occurred in segments of the lake-cargo markets, where the overlap between different producing regions is wide. Producers who gain cost advantages in these markets by forcing freight-rate reductions are amply rewarded as their market boundaries push out into areas formerly controlled by competing districts. Before the new territory is completely captured, however, the original suppliers turn the battleground into a zone of indifference by absorbing prices in a desperate effort to stave off the eventual loss of market area.

In empirical studies, market areas are not so neatly delimited as the simple theoretical models postulate. In addition to boundary lines being blurred by the factors mentioned above, their shapes may be grossly distorted by additional variables. First, products may not be homogeneous. Within the coal industry there are wide ranges of product quality. The fact that coal is a complex bundle of hydrocarbons with different proportions of desirable and undesirable qualities makes subjective valuations difficult. In practice coals can be standardized by valuing them on a cost-per-million-B.t.u. basis.[5] At best this is a crude method of comparison that ignores the true value of the fuel in the use to which it is put.

In actual practice the second theoretical premise that may be violated is the assumption that costs remain constant. This is an area that is dealt with in detail at a later stage in this study.

Finally, assumptions concerning the direction of freight movements and the structure of freight rates may be unrealistic. Transportation routes do not generally proceed along straight lines from points of origin to points of delivery. And the assumption of proportionality in freight-rate making is usually invalid. In an industry like coal mining, in which transportation costs represent such a heavy proportion of the delivered costs, this area needs further study. The section to follow examines the crucial factors in freight-rate construction to clarify the role played by transportation in the delineation of the midwestern coal markets.

The Structure of Freight Rates

In 1959 transportation charges accounted for 42.3 percent of the delivered cost of rail-delivered coal.[6] Clearly, changes in freight-rate relationships can have profound effects on the origin and direction of coal movements.

The proportion of coal originated by railroads has declined in recent years as barge and truck movements have increased. Between 1933 and 1958 the percentage of the national coal production originated by railroads fell from 87.9 to 74.5 percent. The percentage accounted for by water movements meanwhile rose from 3.9 to 10.7 percent, and the percentage of coal moved from the mines to final destination by trucks increased from 4.6 to 12.3 percent.[7] The remaining 2.5 percent in 1958 was consumed at the mine or was shipped by conveyor or pipeline.

Many shipments involve the use of a combination of trans-

[5] The formula for converting coal to a cost per million B.t.u. basis is:

$$\frac{\text{Delivered Coal Cost}}{\text{B.t.u. per pound} \times 2,000 \text{ pounds}}.$$

[6] Schurr, *Energy in the American Economy, 1850–1975*, p. 335.

[7] These data are all secured from Bureau of Mines, *Bituminous Coal and Lignite in 1958*, Mineral Market Summary no. 2974, Sept. 9, 1959, p. 100.

portation media, and the rate structure for each movement is distinctive. There are seven kinds of coal rates that deserve a brief description. Track-delivery rates are used for shipments from the mines to specific consignees located on track sidings. Tidewater rates apply on shipments to ocean ports for eventual movement beyond by ocean-going vessels. These rates traditionally are lower than track-delivery rates, reflecting the lower cost of moving large tonnages from an origination point to a single destination. From the destination tidewater port, coal often moves inland by rail on ex-ocean rates. Thus, coal may move from West Virginia to an inland Massachusetts city on a three-part haul. Total transportation charges would consist of a tidewater rate on coal moved to a port city, an ocean-vessel rate for the water movement to a New England port, and ex-ocean rate on the rail haul inland away from the port city destination. There would be additional transfer charges involved in dumping the coal at a loading pier and in subsequent handling at a New England port.

Coal likewise moves on special rates on shipments to and from inland waterway ports. These rates more directly affect midwest shippers than do ex-ocean and tidewater rates. First, lake-cargo rates are used for rail shipments to Great Lakes' ports for transshipment beyond by lake vessel to other lake ports. Like tidewater rates, lake-cargo rates are lower than those applying on all-rail shipments. The combination of reduced rail rates plus low vessel rates broadens the market for water-borne coal enormously. Corresponding to ex-ocean rates on inland rail shipments away from the destination lake ports are ex-lake rates. Most coal, therefore, goes from mine to destination in the lake market on a two- or three-part haul, moving either by a combination of rail and vessel or via rail-vessel-rail. At each of the transshipping points, the shipper encounters transfer charges that add to the total freight bill.

Ex-river rates apply on shipments that move inland away from river transshipping points after first being transported from the mine to a river terminal by barge. If the mine is not located on a navigable stream, special rates move coal from the landlocked mine to the originating river terminal.

The final type of rail rate combines the features of track-delivery and lake-cargo or tidewater rates. This is the so-called trainload rate. It is used on large shipments, exceeding a prescribed minimum tonnage, when the entire tonnage is consigned to one given destination. The rate per ton on this type of shipment is considerably lower than the single-car rate and permits shippers and carriers to compete with coal delivered by low-cost water transportation and with cheaper competitive fuels.

The principal features of the railroad freight-rate structure on coal movements are:

(1) The wide blanketing of rates for both origins and destinations.[8]

(2) The establishment of differentials among the various competing origin groups.

At an early stage in the development of the American coal industry, the coal regions were subdivided into freight-rate groups to facilitate rate making. Each mine within a group took the same rate to every destination.[9] Originally the freight origin groups were small, but as competition ripened and new mines opened, these groups enlarged. Up to a point it was in the common interest of both operator and carrier to expand an origin district to give the new mine the most favorable freight-rate structure possible.[10]

Destinations were blanketed as well, different-sized areas taking a common rate for shipments from each origin group. As previously noted, this phenomenon has resulted in creating areas of overlap between competing producing districts.

Also at an early stage in the coal industry's development, rate differentials between the various origin groups were created. Generally, groups nearest a given market enjoyed the lowest rate, with more distant groups taking successively higher rates. Often, though, subgroups combined into larger origin districts for par-

[8] "Blanketing" refers to the use of common rates that are applicable over fairly large geographical areas.

[9] There were some exceptions to this statement in the case of short hauls where mileage rates were used.

[10] Kent T. Healy, *The Economics of Transportation in America* (New York: The Ronald Press, 1940), p. 241.

ticular movements. For example, Ohio, western Pennsylvania, and northern and southwestern West Virginia were divided into twenty-two freight districts, each with a different rate structure for shipments into many northeastern United States' markets. But for movements into the midwest, these subdistricts combined into three groups: Ohio, Inner Crescent, an area south of the Ohio

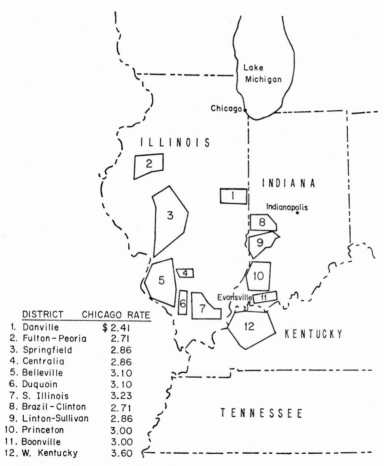

DISTRICT	CHICAGO RATE
1. Danville	$ 2.41
2. Fulton – Peoria	2.71
3. Springfield	2.86
4. Centralia	2.86
5. Belleville	3.10
6. Duquoin	3.10
7. S. Illinois	3.23
8. Brazil – Clinton	2.71
9. Linton-Sullivan	2.86
10. Princeton	3.00
11. Boonville	3.00
12. W. Kentucky	3.60

FIGURE 3. Midwestern freight origin groups (with freight rates to Chicago)

River in eastern Kentucky and West Virginia, and Outer Crescent, located in southern West Virginia and Virginia.[11]

Early competitive relationships divided the midwest into individual origin districts related to each other and to outside districts by prescribed differentials. Figure 3 shows the midwestern freight origin groups in existence today, together with 1960 rail rates to Chicago. Table 1 compares rates from the major mid-

TABLE 1. Rail freight rates per ton from selected origin groups to representative midwestern destinations

Destination	Brazil–Clinton	S. Ill.	W. Ky.	Ohio	Inner Crescent	Outer Crescent
Indianapolis	$1.67	$2.71	$2.76	$4.21	$4.46	$4.56
Chicago	2.71	3.23	3.60	4.69	5.04	5.24
Milwaukee	3.83	4.22	4.58	5.44	5.79	5.99
Minneapolis	4.91	5.23	5.58	7.12	7.47	7.62
Richmond, Ind.	2.88	4.31	4.32	4.21	4.21	N.A.
Oxford, Ohio	3.72	N.A.	N.A.	4.28	3.92	N.A.
Grand Rapids	N.A.	4.80	5.15	4.77	5.27	N.A.
Lansing	N.A.	5.05	5.40	4.42	4.92	N.A.

N.A. = Not available.
Source: Midwest Coal Producers Institute, Inc., *Freight Rate Schedules 2–A, 4, 5, and 7.*

western origin groups with eastern rates to representative midwestern destinations. All figures apply to single-car shipments and are effective June 1, 1960.

Two conclusions clearly emerge from a study of this table. First, the midwest's rate advantage over the east rapidly disappears on shipments to destinations east of a north-south line drawn through eastern Indiana and southwestern Michigan. The southern Illinois district's $.03 disadvantage vis-à-vis Ohio at Grand Rapids widens to a $.63 differential on shipments to Lansing a few miles to the east. The line of demarcation at the Ohio

[11] Thomas C. Campbell, *Bituminous Coal Freight Rate Structure,* Bureau of Business Research, West Virginia Univ., June, 1954, p. 26.

border is even sharper. Brazil–Clinton's advantage over Inner Crescent mines on shipments to Richmond, on the eastern edge of Indiana, amounts to $1.33; yet on movements to Oxford, Ohio, a few miles over the state line, the differential narrows to $.20. This freight structure, however, is a two-edged sword. Effectively limiting penetration of the Ohio and Michigan markets by midwestern producers, at the same time it throws up a barrier protecting the midwest from eastern competition.

The midwest's advantage in its home markets is not as great on a delivered-cost basis as it is on a basis of freight differentials alone. Table 2 expresses this condition in an analysis of freight-rate differentials and delivered-cost differentials in cents per

TABLE 2. Delivered-cost advantage of midwestern coal in its home markets

Destination	Lowest midwest. delv'd. cost[a]	Lowest outside dist. delv'd. cost[a]	Percent midwest. advantage	Lowest midwest. dist. freight rate[b,c]	Lowest outside dist. freight rate[b,c]	Percent midwest. advantage
Chicago, Illinois	30.9	34.2	9.7	$3.21	$4.69	31.6
Indianapolis, Indiana	24.2	32.0	24.4	1.67	4.46	62.5
Richmond, Indiana	29.5	32.3	8.6	2.88	4.21	31.6
Fort Wayne, Indiana	31.2	32.4	3.7	3.28	4.22	22.3
Louisville, Kentucky	21.5	26.7	19.4	1.74	3.02	42.4
Dubuque, Iowa	32.3	39.0	17.2	2.86	4.58	37.5
Ft. Madison, Iowa	30.4	32.6	6.7	2.45	3.17	22.7
Rochester, Minnesota	35.4	40.5	12.6	5.12	4.91	−4.3
Nekoosa, Wisconsin	34.0	38.3	11.2	4.77	5.96	19.9
Kimberly, Wisconsin	34.0	34.6	2.0	4.77	4.90	2.6

[a] Expressed in cents per million B.t.u.
[b] Rates in effect June 1, 1960.
[c] Freight rates for districts with lowest delivered cost.
Sources: Midwest Coal Producers Institute, Inc., *Freight Schedules 2-A, 4, 5, 6, 7*, June 1, 1960; Appendix, Table 43.

million B.t.u. for ten representative midwestern consuming points. Delivered-cost differentials range from 2 percent at Nekoosa, Wisconsin, to 24.4 percent at Indianapolis, deep in the midwest's home territory. The differentials deepen even more for markets adjacent to the midwestern mine fields.

To most of the destinations listed in Table 2 the midwest's freight-rate advantage expressed in percentages is two to three times greater than its percentage advantage in delivered costs. But the ability of outside district producers to reduce mine prices sufficiently to overcome delivered-cost disadvantages is severely limited. Since mine prices represent approximately 60 percent of total delivered prices, outside district operators, on the average, must reduce mine prices 10 percent to eliminate a 6 percent delivered-price disadvantage in the absence of interdistrict freight differential changes. In an industry operating under narrow average profit margins, it is difficult to overcome most of the delivered-cost differentials listed in Table 2 through reduced mine prices. But even if outside district operators cut prices to overcome the existing differentials, midwestern operators can presumably protect their market positions by matching their competitors' reductions.

That delivered-cost differentials favoring the midwest are lower than freight-rate differentials indicates the relatively lower quality of midwestern coal compared with coal in the east. It also forcefully demonstrates the market protective value of the existing freight-rate structure—an outgrowth of the existing location of coal reserves.

The implications of this rate structure are enormous. It creates a vast island containing some of the country's largest coal markets which are virtually captive to midwestern producers. Eastern rail shipments to utility and industrial consumers can effectively penetrate only the periphery of this protected area. Shipments into the interior of this region to retailers and metallurgical coal users occur, but in these markets competition between eastern and midwestern shippers is minimized or nonexistent. In the largest segment of this isolated area, therefore, midwestern producers vie only with each other for a share of the market. The ramifications of this insularity as evidenced in the conduct and performance of midwestern producers are obvious. Interregional competition is severely limited.

A coal producer ordinarily encounters fuel competition on three fronts. First, he competes directly with other producers operating in the same coal region. Additionally, he faces competition from operators in outside fields. Finally, he competes with alternate fuels in markets in which fuels are used interchangeably. In the latter case gas unavailability and high fuel oil costs limit interfuel competition in many industrial and utility situations. The relative isolation of interior midwestern markets resulting from the freight-rate structure outlined above damps down competitive pressures from outside coal fields. Thus, in a large portion of the midwestern market area, midwestern producers compete only among themselves. This phenomenon, coupled with the increasing concentration of midwestern production to be discussed in Chapter IV, significantly influences the behavior of the midwestern industry.

The second conclusion to be drawn from the rate differentials exhibited in Table 1 is the relative improvement in the rate structures for western Kentucky and southern Illinois vis-à-vis Brazil–Clinton on shipments farthest removed from the origin districts. Brazil–Clinton's $1.09 advantage over western Kentucky on shipments to Indianapolis narrows to $.67 on movements to Minneapolis. Narrowing of differentials between Brazil–Clinton and southern Illinois is even greater, the latter's freight disadvantage being reduced from $1.04 on Indianapolis shipments to $.32 on hauls to Minneapolis. This freight-rate structure combines with the inherent superiority of southern Illinois and western Kentucky coals (plus lower operating costs for mines in the latter district) to foreclose Indiana coal from most remote markets.[12] There are compensating advantages for Indiana shippers, however, since

[12] Even if freight differentials between western Kentucky and southern Illinois on the one hand and Indiana on the other do not narrow, Indiana is increasingly worse off relatively in distant markets on account of its lower quality. Assume a delivered price of $5.50 a ton for a shipment to a market near the Indiana coal field. Producing coal with a heat content of 11,500 B.t.u., Indiana producers suffer a disadvantage of 4.78 cents per hundred B.t.u. when competing with higher-quality coals. In more distant markets, commanding higher delivered prices, the disadvantage increases. Thus, if the delivered price is $8.50, the value differential rises to 7.39 cents per hundred B.t.u.

they possess a near-monopoly power in most of the markets within their home state.

The results of the practice of establishing rail freight rates that are nonproportional with respect to distance hauled are indicated in Table 1. High terminal and fixed charges characterize railroad cost structures. Short hauls, therefore, absorb proportionally more of these charges in their costs than do longer hauls, and this fact reflects itself in the different level of rates for hauls of different lengths. Average rail rates on 1953 movements from Illinois, Indiana, and western Kentucky mines to midwestern destinations ranged from 7 to 15.3 mills per ton-mile, with the rates inversely related to the distances hauled, which ran from 104 to 858 miles.[13]

River movements by barge are considerably cheaper than rail transportation. Freed from expensive right-of-way construction and maintenance costs and enjoying lower terminal charges than railroads bear, barges can transport coal at rates averaging 3 to 4 mills per ton-mile. Thus, barge rates per ton-mile are not only lower than rail charges but there is less spread between the minimum and maximum than there is with rail rates. Since barges must follow the course of inland waterways, the movement of barge coal from mine to market follows a circuitous route. Still, the lower rates per ton-mile usually more than compensate for the extra distance barges must cover, so that consumers on navigable rivers prefer barge- to rail-delivered coal.

The comparative rate structures of barge and rail movements help define the market areas available to each mining district. Easy access to river-loading terminals becomes imperative if a district intends to move coal beyond the limits imposed by high rail rates. Inability of most Indiana mines to secure low rail rates to Ohio River ports prevents their competing effectively in river markets with the western Kentucky field which has ready access both to the Tennessee and the Ohio Rivers. The economy of barge transportation permits western Kentucky operators to move coal as far away as Tampa, Florida, on all water movements via the Green, Ohio, and Mississippi Rivers and the Intercoastal Waterway system. Furthermore, the advantages of plentiful water,

[13] Osmond L. Harline, "Economics of the Indiana Coal Mining Industry" (Unpubl. diss., Indiana Univ., 1958), p. 354.

cheap transportation, and low-cost barge coal attracts industry to inland waterways, further strengthening the market position of mines accessible to river terminals.

Intermediate in cost between rail and barge movements are lake-vessel rates. Their relative cheapness permits eastern coal to move into Lake Michigan and Lake Superior markets in competition with midwestern mines when exorbitant all-rail rates would ordinarily exclude it. Their coals unacceptable to Wisconsin buyers and lacking suitable port facilities, midwestern operators for many years were excluded from lake markets that should have fallen within their orbit. With the construction of adequate transfer facilities at South Chicago after World War II and the redesign of boiler equipment to permit the use of lower-rank coals, the midwest recently has displaced eastern shippers in many of the lake markets. Whatever economic advantage eastern dock coal possesses in the port cities it soon loses when coal is transshipped inland. Beyond a fairly short radius ex-dock coal cannot effectively compete with midwestern all-rail coal.

For very short movements truck transportation is cheaper than either rail or water transportation. It is particularly useful for hauls to off-track industries which, in the absence of truck shipments direct from the mine, must receive coal that is twice handled—first by the originating carrier and then by a delivering trucker who secures it at an intermediate storage location. For most movements to plants capable of receiving direct rail or barge shipments, truck transportation is uneconomical beyond a radius of fifty to sixty miles from the mine.

Table 3 presents a breakdown of shipments from midwestern mines in 1959 by the various methods of transportation described above.[14] The figures distort somewhat the proportion of midwestern production transported by rail. Much of the river tonnage and all of the coal moved over the lakes originate by rail. Indeed, if the figures are broken down by revenues earned, railroads would account for perhaps 75 percent of the total. But the trend in mid-

[14] The discussion has ignored coal moved by conveyor or pipeline. In several instances coal is moved from the mine to adjacent or nearby electric utilities by rubber belt conveyors, but these movements represent a small proportion of total coal output. Shipments by pipelines are equally insignificant in terms of coal tonnage transported, but their potential use is vastly greater.

TABLE 3.   Distribution of midwestern coal, by method of transportation, 1959

Method of movement	Tons (thousands)	Percent
All-rail	48,019	53.2
River and ex-river	27,784	30.7
Great Lakes	6,730	7.5
Truck, private railroad and conveyor	7,822	8.6
Total	90,355	100.0

Source: Bureau of Mines, *Bituminous Coal and Lignite Distribution, 1959*, Mineral Market Report, M.M.S. no. 3035.

western coal transportation increasingly is away from the railroads, with barge movements assuming greater importance.

Isolation of Midwestern Market Area

The previous section explains the structure of freight rates and stresses the importance of freight rates in determining market areas by affecting coal's delivered cost. This section will carry the analysis further, defining more explicitly the shape of the midwestern industry's market area. The extent of interregional coal competition at various points within the market area will be explored. Furthermore, the uniqueness of the midwest's sheltered market position will be compared with a typical outside district shipping to fiercely competitive markets that are traditionally associated with the coal industry.

Since 1959, the Bureau of Mines has assembled distribution data, breaking down shipments from the various producing districts to each consuming state. Table 4 presents this data for 1959 for the six principal states in the midwest's market area. Retail deliveries from outside districts are excluded since competition between high- and low-quality coal shippers is minimal.[15] Domestic consumers ordinarily purchase either a high-price, quality coal or a low-price, lower-rank coal. If the former choice is

[15] The word "outside" used in this context refers to coal districts outside of the Eastern Interior region.

made, the consumer stresses cleanliness and ash content over price; if a lower-rank coal is chosen, price is the guiding consideration. In the midwestern states, this amounts to a choice between eastern and midwestern coals. As a result, competition is among shippers of equal-rank coal and not among coal regions producing coals of unequal rank.

TABLE 4. Distribution of coal to selected states in midwestern market area, 1959 (thousand tons)

To \ From	Indiana	Illinois	W. Ky.	Total midwest	Other states	Total
Indiana	11,707	1,243	2,890[a]	15,840	2,033	17,873
Illinois	1,565	26,081	6,609	34,255	506	34,761
Wisconsin	596	4,263	2,221	7,080	3,357	10,437
Iowa	100	2,732	401	3,233	1,426	4,659
Missouri	0	4,013	189	4,202	2,235	6,437
Kentucky	0	3,370	5,230	8,600	810	9,410
Total	13,968	41,702	17,540	73,210	10,367	83,577

[a] Excludes estimated 2 million tons of western Kentucky coal shipped to Ohio utilities but listed in the Bureau of Mines' report under Indiana shipments.

Source: Bureau of Mines, *Bituminous Coal and Lignite Distribution, 1959,* Mineral Market Report, M.M.S. no. 3035.

Likewise, Table 4 excludes consideration of all shipments of metallurgical coal from eastern mines to coke and gas plants. Being put to a specialized use, coal of this type does not compete with midwestern steam coal, except in the few plants which in 1959 used 696,000 tons of marginal southern Illinois coking coal. This tonnage is excluded as well in the table.

The remaining tonnage, therefore, covers shipments into markets within the six states where the various coal regions competed directly with one another. The midwestern mines dominated these markets, accounting for 87.6 percent of the 1959 shipments into them. The extent of this domination varied markedly. Eastern mines accounted for nearly one third of the coal shipped to Wisconsin. Similarly, a third of the coal shipped to Missouri and Iowa came from districts outside the midwest, principally from mines in these two states as well as from Kansas and Oklahoma mines. Understandably, the dominance of midwestern shippers was much stronger in their home states than it was elsewhere. Midwestern mines shipped 86.5 percent of the tonnage sold in Indiana, 91.6 percent in Kentucky, and 98.5 percent in Illinois.

Table 4 accounts for 81.6 percent of midwestern production. The eight other states absorbing most of the remaining midwestern tonnage relied less heavily on midwestern shipments than did the six base states.[16] In 1959, these states took 15,123,000 tons of midwestern coal, only 18.3 percent of their total requirements.[17] Only in Minnesota, among the eight states, did the midwest's market share exceed 50 percent; there it accounted for 57 percent of the total consumption.

It is illuminating to narrow the focus from a study of market dominance in all industrial and utility markets just to the electric utility market. This is done for two reasons: first, comparable data are available for two time periods to provide a rough indication of trend, and second, consumption of coal by electric utilities is growing at such a fast rate relative to other use categories that this market increasingly dominates the total fuel market.

Tables 5 and 6 present comparable electric utility consumption data for the six-state base market area for 1928 and 1959.[18] Midwestern mines supplied 87.8 percent of the coal burned by the utilities in these states in 1959, contrasted with 78 percent in

[16] These eight states are: Minnesota, Ohio, Michigan, Georgia, Florida, Tennessee, Alabama, and Mississippi. The Bureau of Mines' distribution reports lump together shipments to Georgia and Florida and to Alabama and Mississippi in two figures. There is no way of segregating the tonnages by states in these cases. Most, if not all, of the midwestern shipments to these two groups of states move into Florida and Alabama.

[17] Again, retail and metallurgical shipments were excluded.

[18] Actually, only the 1928 data are given in terms of consumption. The 1959 figures represent shipments and vary from consumption totals by changes in inventories, including coal in cars en route to destination. The differences between consumption and shipments are slight. In any event they cannot appreciably alter the percentage relationships between producing districts and between the two time periods, although absolute tonnage figures may be slightly distorted.

TABLE 5. Consumption of coal in electric utility plants, selected states, 1928 (tonnages in thousands)

| Consuming state | Producing district | | | | | |
| | Midwest | | Other states | | Total | |
	Tons	Percent	Tons	Percent	Tons	Percent
Indiana	2,099	87.1	315	12.9	2,414	100.0
Illinois	6,074	99.2	46	10.8	6,120	100.0
Wisconsin	85	10.0	764	90.0	849	100.0
Iowa	409	46.7	466	53.3	875	100.0
Missouri	399	33.2	801	66.8	1,200	100.0
Kentucky	295	53.8	253	46.2	548	100.0
Total	9,361	78.0	2,645	22.0	12,006	100.0

Source: Wm. H. Young, *Sources of Coal and Types of Stokers and Burners Used by Electric Public Utility Power Plants,* pp. 74–77.

1928. Their share of the market increased in each state, but the most striking increases were in Wisconsin and Kentucky. Midwestern shipments rose from 10 to 76.2 percent of the total Wisconsin utility requirements in response to the opening of Lake

TABLE 6. Shipments of coal to electric utility plants, selected states, 1959 (tonnages in thousands)

| Consuming state | Producing district | | | | | |
| | Midwest | | Other states | | Total | |
	Tons	Percent	Tons	Percent	Tons	Percent
Indiana	10,284	90.1	1,158	9.9	11,442	100.0
Illinois	17,941	99.6	125	0.4	18,066	100.0
Wisconsin	4,361	76.2	1,360	23.8	5,721	100.0
Iowa	1,134	54.5	946	45.5	2,080	100.0
Missouri	1,343	43.7	1,734	56.3	3,077	100.0
Kentucky	6,880	92.9	544	7.1	7,424	100.0
Total	41,943	87.8	5,867	12.2	47,810	100.0

Source: Bureau of Mines, *Bituminous Coal and Lignite Distribution, 1959,* Mineral Market Report, M.M.S. no. 3035.

Michigan to midwestern lake-cargo coal. Western Pennsylvania mines' share fell from 83 percent to less than one percent during the thirty-one-year period. In Kentucky the midwest's market share increased from 53.8 to 92.9 percent as new electric utility construction was attracted to the Green River and western Ohio River areas accessible to low-cost western Kentucky and southern Illinois coals.

As useful as the current state distribution figures are, however, they fail to convey the true extent of market dominance attained by midwestern shippers in the areas in which midwestern coal is consumed. The data distort the picture by using artificial political boundaries to define market areas, instead of using more rational economic boundaries. The distortions are numerous. The Wisconsin shipments originating in eastern districts move into a narrow geographical area bordering Lake Michigan. Much of the remainder of the state is virtually captive to the midwest. The same situation prevails in Minnesota where midwestern shippers supply only 55 percent of the state's coal requirements.[19] Here also, eastern competition centers around the port cities, the rest of the state's markets being dominated by the midwest. Little of the outside coal shipped into Missouri and Iowa competes directly with midwestern tonnage. Interregional competition exists in the areas where the natural markets of the midwestern mines on the one hand, and Missouri and Iowa mines on the other, overlap; in the sections of the states where midwestern districts have a commanding freight-rate advantage, the midwest enjoys a limited monopoly position.

An effort was made to define, more precisely than the Bureau of Mines' data permit, the market area dominated by Illinois, Indiana, and western Kentucky mines. Figure 4 graphically portrays the limits of their control. The heavy dark line encloses a region within which at least one midwestern mining district commands a delivered-cost advantage over the lowest-cost, outside producing district. An appendix describes in detail the calculations necessary to derive the boundary line, but a word is called for here to explain the general method used.

[19] Bureau of Mines, *Bituminous Coal and Lignite Distribution, 1959,* Mineral Market Report, M.M.S. no. 3035, March, 1960, p. 17.

FIGURE 4. Area of midwestern market dominance

It was first assumed that coal buying was rational, i.e., that purchases were made on the basis of the lowest cost per million B.t.u. This assumption limited the analysis primarily to the electric utility and industrial market segments and excluded retail coal, much of which is purchased "irrationally." [20] Additionally, it was assumed that the supply curves for each district were sufficiently elastic to permit the shipments called for under the distribution arrangements implied by the creation of the hypothetical market boundary. This is not an unduly restrictive assumption. Variations in delivered prices for shipments from the various midwestern districts are slight in many instances. In these cases any one of several districts could have supplied coal at prices below the level of the lowest outside district. If, despite this flexibility, the distribution arrangement assigned more tonnage to a given district than it currently produces, the difference could be made up in the short run by utilizing redundant capacity. If still more tonnage were required, the district could build additional capacity.

Knowing the states in which midwestern coal is presently burned and studying the freight-rate structure permitted the delineation of the midwestern market area in a rough fashion. Delivered costs per million B.t.u. were computed for hypothetical deliveries to eighty-six cities in the eight states where the boundary was assumed to lie. In each case the lowest-cost means of transportation was selected for shipments from each district. Rail rates ordinarily were used for comparisons to destinations along the eastern, southern, and western periphery of the market area. In the lake region, eastern rail-lake rates were used for movements to port cities and rail-lake-rail rates for shipments to interior Wisconsin and Minnesota destinations. To the same destination, midwestern delivered costs were computed using either rail-lake rates or all-rail rates, whichever was cheaper.

Several districts from outside the midwestern region provided the lowest-cost competition for the midwest at various points along the boundary. Ohio and Inner Crescent district mines competed most strongly along the eastern border and in the lake

---

[20] A careful analysis of Tables 12 and 13 reveals that much industrial fuel oil is purchased at prices in excess of average coal prices. Doubtless, some of these purchases are irrational; however, in some instances economies of superintendence, maintenance, furnace efficiency, and so forth, resulting from the use of fuel oil, overcome higher costs of the fuel itself. In any event, nearly 90 percent of the industrial fuel requirements of the states listed in Tables 12 and 13 were filled by lower-cost coal and gas.

territory. In the south, comparisons were made with eastern Kentucky and Jellico, Tennessee, mines. Missouri mines provided the strongest competition in their home state as well as in Iowa and southern Minnesota.

Throughout the length of most of the boundary, the midwest's cost advantage increased rapidly inward, away from the perimeter. For example, at Marshalltown, Iowa, the lowest-cost midwestern coal delivered for two mills per million B.t.u. more than Missouri coal (less than a one percent difference); yet, at Muscatine, Iowa, midwestern coal had a 20 percent delivered-price advantage. At points closer to the center of the market area, midwestern producers command an even greater advantage.

It remains to be seen whether empirical data can verify the accuracy of the hypothetical boundary set out in Figure 4. Fortunately, distribution data are available for 1945–1946 which throw light on the issue. The Bureau of Mines made a breakdown of tonnages shipped by different producing districts to the various market areas that were created by the Bituminous Coal Commission. Illinois, for example, was divided into eight market areas for reporting purposes, instead of just one, as is done with current distribution statistics. This kind of breakdown permits a more detailed analysis of the concentration of sales of any producing district in different parts of the midwestern market.

Figure 5 graphically summarizes the extent of the midwestern industry's market dominance in this area. Specifically, the area has been divided into regions where midwestern shippers supplied 50 to 75 percent, 75 to 95 percent, and over 95 percent of the total electric utility and industrial coal requirements.[21] The areas in which midwestern producers capture over 95 percent of the market tonnage correspond reasonably well to the area included within the hypothetical boundary in Figure 4. If the 75 to 95

[21] Including Wisconsin and southeastern Minnesota in the 95 percent concentration zone is somewhat misleading. The available statistics divide tonnages into two categories: all-rail shipments and coal shipped over the Great Lakes. Only the all-rail shipments are broken down by originating districts, and the percentage of market dominance is measured here using these figures. If lakedock tonnages were included, the midwest's share of these markets would be sharply lower. However, since freight-rate relationships have not altered appreciably since the war, the 95 percent line would probably closely approximate the hypothetical boundary running through these states as shown in Figure 4.

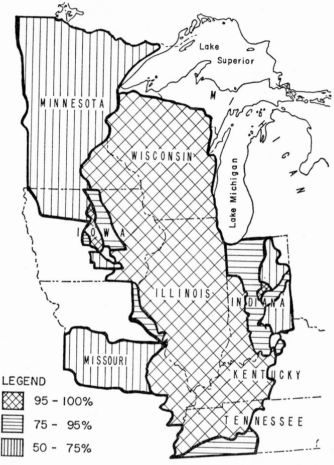

FIGURE 5. Areas of varying degrees of midwestern market dominance

percent market share areas are included the correspondence is even closer. The only major area in which the 1946 distribution differs sharply from the hypothetical area is in that part of Minnesota outside the southeastern section of the state. And here differences probably are due more to data limitations than to actual deviations from the hypothetical boundary.

Electric utilities and industrial users in the +95 percent zones purchased 30,688,463 tons of midwestern coal in the year ending March, 1946, or 74.1 percent of midwestern shipments to these consumer categories.[22] Applying the same percentage to 1959 shipments indicates that nearly 59,000,000 tons of midwestern coal consumed by these users moved into sheltered market areas where interregional competition was limited or nonexistent. If shipments to retailers and American railroads were included, 75.9 percent of all the midwestern industry's 1959 tonnage was protected from outside competition.

Nothing has happened in the period between 1946 and 1959 to cast suspicion on the interpolation made here. If anything, the dominance of the midwest in its base markets has probably increased. For example, in 1945–1946 midwestern shippers accounted for 94.1 percent of Illinois and Indiana industrial and electric utility shipments, whereas by 1959 their share of these markets had increased to 95.2 percent. The areas in eastern Indiana where the midwestern market share in 1946 was limited to the 50 to 75 percent level probably accounted for most of the increase in the midwest's domination in 1959.

Compare the midwestern industry's market share position with that achieved by other coal regions. The largest coal district in the country is District 8, a producing region comprising eastern Kentucky, parts of West Virginia, Tennessee, and Virginia.[23] Table 7 compares District 8's share of the electric utility and industrial markets in the states where its coal is principally consumed with the use of coal from competing districts. The nine states listed account for 85 percent of District 8's sales to these

---

[22] Bureau of Mines, *Bituminous Coal Distribution by Market Areas, Calendar Year 1944, Year Ended September 30, 1945 and Coal Year 1945–46,* Mineral Market Report, M.M.S. no. 1500, March, 1947. This percentage excludes midwestern coal shipped to Wisconsin and southeastern Minnesota.

[23] Under the Bituminous Coal Act of 1937 (commonly referred to as the Guffey Act) the industry was divided into 23 coal districts for administrative purposes. Within the coal industry reference is often made to these numbers when identifying a particular district. In some cases the district numbers cut across state lines; in others they refer to all the coal produced within a single state; in still other cases districts comprise only a part of one state. In the midwest, Indiana is District 11, Illinois, District 10, and western Kentucky, District 9. The rest of Kentucky is included within District 8.

TABLE 7. Distribution of District 8 coal to electric utilities and industrial plants, selected states, 1959 (tonnages in thousands)

Destination	District 8	Percent	Other districts	Percent	Total	Percent
Michigan	9,264	47.0	10,586	53.0	19,850	100.0
Virginia	8,545	86.0	1,416	14.0	9,961	100.0
N. Carolina	7,525	95.0	436	5.0	7,961	100.0
Ohio	7,026	20.0	28,467	80.0	35,493	100.0
Tennessee	6,255	49.0	6,371	51.0	12,626	100.0
W. Virginia	4,008	42.0	5,394	58.0	9,402	100.0
S. Carolina	3,083	98.0	52	2.0	3,135	100.0
Wisconsin	2,214	31.0	7,765	69.0	9,979	100.0
Indiana	1,632	14.0	9,810	86.0	11,442	100.0
Total	49,552	41.5	70,297	58.5	119,849	100.0

Source: Bureau of Mines, *Bituminous Coal and Lignite Distribution, 1959*, Mineral Market Report, M.M.S. no. 3035.

consumer categories. In these states this district accounts for only 41.5 percent of utility and industrial sales, less than half the share earned by the midwest in its principal market area. Competition within District 8 is more severe than in the midwest in another important respect. The 49,500,000 tons shown in Table 7 represent only 55 percent of the district's 1959 total production. Most of the remaining tonnage includes electric utility and industrial coal shipped to other states and large quantities of coal shipped to retail, export, and metallurgical coal markets. Some of the latter shipments are captively controlled by major steel manufacturers and as a result do not move into normal market channels. But the remaining metallurgical coal, plus all of the export and retail coal, must perforce compete directly with coal from a number of outside districts.

Even in the market areas where District 8 possesses a regional monopoly—in North Carolina and South Carolina, for example—intradistrict competition is rather severe. In 1957, 204 producing companies in District 8 producing over 100,000 tons accounted for approximately 95 percent of the district's production. An additional 127 companies, each mining between 10,000 and 100,000 tons, and countless hundreds of "dog hole" operators mining less than 10,000 tons produced the remaining tonnage.[24] Whatever monopoly powers District 8 as a whole might enjoy from certain locational advantages is vitiated by the pressure from so many intradistrict competitors. This condition sharply contrasts with that faced by midwestern operators where output centers in a few hands. Chapter IV elucidates this point in detail.

The pattern of interregional competition facing District 8 illustrates the condition existing in most of the coal markets outside the midwest. There are exceptions, but the tonnages involved in these exceptions are comparatively small. A coal region identified in the trade as District 15, comprising mines in Oklahoma, Kansas, and Missouri is sufficiently isolated geographically to form a regional monopoly. But this area accounts for barely one percent of the nation's coal production. The other major producing regions, Ohio, Pennsylvania, and northern West Virginia, encounter stiff interdistrict competition in all markets outside of circumscribed areas near each district's coal fields, just as District 8 does.

Comparison with Henderson's Markets

The publication of Henderson's linear programming study of the coal industry requires that we add a footnote to our analysis of the midwestern submarket.[25] The reader who is familiar with Henderson's work may be disturbed that the markets here delineated differ markedly from those used in his "efficient" solutions. The discrepancies demand explanation.

Henderson seeks to determine the most efficient delivery levels for the bituminous coal production for 1947, 1949, and 1951, using the so-called "transportation problem" as a framework. He uses actual consumption levels to indicate "demand" (assumed to be infinitely elastic with respect to price); capacity is subdivided into its strip and underground components. He combines states into fourteen consumption districts, eleven of which are also production areas. These historical consumption (demand) levels and

---

[24] *Keystone Coal Buyer's Manual* (New York: McGraw-Hill Publishing Co., 1958).
[25] Henderson, *The Efficiency of the Coal Industry.*

capacities together with strip and underground unit costs, averaged for each district, and average transportation costs from the producing districts to each consuming district comprise the data used to implement the model. "The efficient solutions give the delivery levels which would have prevailed if total costs had been minimized in (the three) years."[26] Henderson compares his least-cost solutions with actual delivery levels to determine the extent of inefficiency in the industry.

Henderson's analysis is not directly comparable with ours since he divides the midwest into two segments, putting Indiana and Illinois (along with Michigan) into one production-consumption district (Henderson's District 6) and Kentucky into another, combining it with Virginia and Washington, D.C. (these three areas make up Henderson's District 3). Data limitations prevent dividing Kentucky into its eastern and western components, and Henderson acknowledges this deficiency in the study. It is a deficiency which casts doubt on the model's results, particularly when it is combined with its other weaknesses.[27]

The so-called "efficient" solutions assign the entire states of Indiana, Illinois, and Michigan to the Illinois–Indiana market area for 1949 and 1951. In 1947 District 3 shares a part of this market, presumably shipping into Michigan. Despite the proximity of northern Illinois mines to eastern Iowa markets and of Belleville, Illinois, district mines to the big St. Louis market, the model assigns no tonnage from District 6 into District 7 (Iowa, Missouri, Kansas, Arkansas, Oklahoma, and Texas) in 1947. Instead, this market is served partly by mines within the district and mostly from West Virginia mines! Moreover, District 6 ships nothing into the states of Wisconsin and Minnesota, sharply contrasting with our analysis which places large sections of these states within the orbit of the midwestern mines.

How can we account for these apparently bizarre allocations of output? Henderson correctly recognizes that "an individual area is not necessarily supplied by the deposit which is least costly if

[26] Henderson, p. 41.

[27] For a penetrating criticism of Henderson's work (plus justifiable praise as well) see Mark Nerlove's "On the Efficiency of the Coal Industry," *Journal of Business*, July, 1959.

that area is considered in isolation from the others." [28] Nevertheless, it is hard to ignore the fact that the model grinds out solutions calling for competing mining districts to crosshaul into each
other's natural markets, when a rearrangement of the coal distribution would put each district into nearby markets. Blame for
the strange solutions rests in the model's underlying framework.
It labors under two major handicaps. First, it tries to combine
into presumably representative figures averages of unit costs, coal
qualities, and transportation charges for each of the twenty-two
strip and underground deposits. This procedure is questionable. If
a district's strip mines have lower costs on the average than its
underground mines, the model assumes that *all* of the district's
strip mines possess lower costs than *all* of its underground mines!
From available cost records we know that this assumption is false.
Second, the model errs in assigning transportation rates from
each district to specific states on the basis of historical movements
to those states. Thus, if Pennsylvania has favorable delivered
costs on combined rail-lake movements to Wisconsin and Minnesota port cities, the model assigns the same average transportation
charges for shipments to *all* destinations within the two states.
The current failure of Pennsylvania to supply the whole of these
states does not necessarily involve an inefficient allocation of resources, as Henderson's analysis implies; rather, it may stem
merely from the existence of transportation rates too high to
permit Pennsylvania to compete with the midwest in those areas
in the states which Pennsylvania does not now serve. Thus Henderson's analysis errs by gathering transportation data by states
when in fact natural coal markets disregard artificial state boundary lines. [29]

But the analysis has another shortcoming, and because of it
the model's solutions are not unique. Henderson's measures of
inefficiency and his assignment of markets to the various producing districts are valid only for a model with the consumption
districts he has arbitrarily formed. For example, grouping Michigan with another district and not with Illinois and Indiana
changes the solutions, as would any of the innumerable other re-

---

[28] Henderson, p. 9.
[29] He had no alternative, however, since his basic data were grouped by states.

arrangements of consumption districts. Thus, if Wisconsin replaced Michigan in forming District 6, the model's solution doubtless would call for Illinois and Indiana now to supply Wisconsin, leaving Michigan to be supplied by an eastern district — a more reasonable solution, by the way. These substantial changes in the areas comprising regional submarkets would flow not from changes in the underlying cost-and-coal-quality data but, instead, from an arbitrary recomposition of the consumption districts. Again, the averaging process lies at the heart of the problem. Assigning the Michigan market to the Illinois–Indiana producing districts results from the invalid assumption that average transportation charges from Illinois and Indiana mines to destinations in those two states plus Michigan give a true picture of transportation charges to *Michigan itself*.

This extended criticism of Henderson's work should not obscure the great promise which the linear programming technique offers for the solutions of problems similar to the one Henderson attacks, nor should it detract from the high quality which characterizes most of his analysis. The critique challenges the shaping of midwestern markets provided in the Henderson model. Lest we be accused of attacking a straw man, we admit that even if the model were properly constructed, its resulting market areas would be designed to effect an efficient solution rather than to represent so-called "natural" market areas. Nevertheless, though they might differ from the natural market areas, the differences would be insignificant. Important for this discussion is the fact that the differences between Henderson's midwestern markets and those resulting from true, least-cost considerations are far from insignificant.

CHAPTER III

The Demand for Coal

Coal is useful principally as a source of energy. In the combustion process, stored energy is released as heat that is used either to warm air or to convert water to steam. The steam is useful directly in production processes as an energy source to power turbines or other industrial equipment.

Consumer Uses

As a heat and energy source, coal is consumed primarily by the electric utility industry, in domestic heating, by manufacturing industries in general, and, to a declining extent, by the railroad industry. As a base for the production of metallurgical coke, coal is also used by the steel and foundry industries and enters the export market.

Figure 6 indicates the distribution in the United States among these major use categories for the period 1933–1961. Several features stand out in the analysis of these figures. First, the over-all demand for coal has been secularly stagnant. Indeed, although these data do not show it, the beginning of the stagnation period can be traced to the mid-1920's. The only relief from this condition of stationary secular demand came during the forced draft years of World War II and for several years thereafter.

The second significant characteristic of these data is the shift in relative importance among several consumer categories. Consumption by railroads declined from a high of 132 million tons in 1944 to less than 2 million tons in 1961. Sales to retailers during the same period fell 77 percent, as coal lost out in the domestic market to competitive fuels. Gains in consumption by the electric utility industry have partially counteracted these losses. Growing

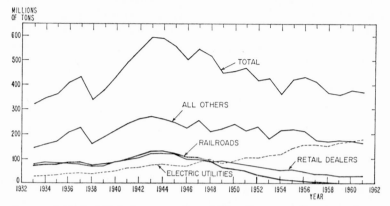

FIGURE 6. United States' consumption of bituminous coal (by class of consumer)

at a rate of approximately 7 percent a year, consumption rose from 27 to 180 million tons from 1933 to 1961.

Tables 8 and 9 present the distribution of coal according to major use categories. The data compare the midwest with the rest of the industry.

TABLE 8. Distribution of bituminous coal among major consumer classes from mines in Illinois, Indiana, and western Kentucky, 1959 (tonnages in thousands)

Consumer	Western Kentucky		Illinois		Indiana		Total	
	Tons	Per-cent	Tons	Per-cent	Tons	Per-cent	Tons	Per-cent
Electric utilities	21,601	72.2	24,410	53.4	8,499	57.7	54,510	60.3
Coke plants	5	0.0	696	1.5	0	0.0	701	0.8
Retail	3,034	10.1	5,135	11.2	1,032	7.0	9,201	10.2
Railroad fuel	247	0.8	520	1.1	191	1.3	958	1.1
All others[a]	5,046	16.9	14,994	32.8	4,971	34.0	25,011	27.6
Total[b]	29,933	100.0	45,755	100.0	14,693	100.0	90,381	100.0

[a] Includes mostly shipments to industrial plants.

[b] Shipments from the midwest accounted for 21.9 percent of the national bituminous coal production, 7.3 percent coming from western Kentucky, 11.1 percent from Illinois, and 3.5 percent from Indiana.

Source: Bureau of Mines, *Bituminous Coal and Lignite Distribution, 1959*, Mineral Market Report, M.M.S. no. 3035, March, 1960.

TABLE 9. Distribution of bituminous coal and lignite among major consumer classes from districts outside of midwest, 1959 (tonnages in thousands)

Consumer	Tons	Percent
Electric utilities	111,490	34.7
Coke plants	76,299	23.7
Retail	19,799	6.1
Railroad fuel	2,042	0.6
All others	74,989	23.4
Export	37,000	11.5
Total	321,619	100.0

Sources: Bureau of Mines, *Minerals Yearbook, 1959;* Bureau of Mines, *Bituminous Coal and Lignite Distribution, 1959,* Mineral Market Report, M.M.S. no. 3035, March, 1960.

Significantly, the midwest relies most heavily on shipments to that segment of the market which is growing fastest — the electric utility industry. This feature of its market structure gives the midwest a dynamism that is absent in the rest of the industry. It also provides midwestern operators with a demand for their product more stable than the other districts enjoy.

Demand for coal from coke plants and from export markets is especially volatile. Districts outside the midwest count on these markets to consume a substantial portion of their annual output. These markets together in 1959 accounted for 35.2 percent of their total production. Table 10 indicates the fluctuating character of demand in the coke and export markets, in contrast to the steadily increasing trend of demand by the electric utility industry.

Consumption by coke and gas plants increased in six years, declined in four. The average annual change was 12.7 percent, and for the entire period, consumption fell 8 percent. Demand generally moved cyclically in response to changes in the Gross National Product, but the amplitude of its cyclical phases exceeded those of the G.N.P. fluctuations. The volatility of steel demand, from which the demand for coke is largely derived, accounted for these wide swings.

TABLE 10. Percentage change in consumption of United States' coal by electric utility industry, coke and gas plants, and export markets, 1950–1959

	Percent change over previous year's consumption									
	1950	1951	1952	1953	1954	1955	1956	1957	1958	1959
Coke and gas plants	+14	+9	−14	−5	−17	+26	−2	+2	−29	+9
Export[a]	−79	+1336	−21	−47	+7	+126	+41	+21	−34	−35
Electric utility	+9	+15	+1	+9	+3	+22	+10	+2	−3	+9

[a] The export figures represent shipments from the United States and do not necessarily coincide with consumption of United States' coal abroad.
Source: Bureau of Mines, *Minerals Yearbook, 1959.*

Export demand oscillated even more. Annual changes in demand for the period covered in Table 10 ranged from increases of 1,336 percent to reductions of 47 percent, with the over-all demand down 10.8 percent. General economic and political conditions abroad, availability of foreign fuel resources, and such transient conditions as the 1956 Suez crisis contributed to the violent shifts in demand in this market segment. Though some export coal is sold on long-term commitments, much of it is purchased on a spot basis, with prices reflecting the short-term forces of supply and demand. Certain Appalachian coal regions that rely heavily on export sales move swiftly from periods of relative prosperity to deep depression depending on the state of the export market.

Fortunately for midwestern producers, sales to electric utilities, their major market, have shown less violent fluctuations and, in addition, generally have increased. Only in the recession year of 1958 did sales to this group fall off, and then only 3 percent. Each of the other years registered increases, adding up to a 107 percent gain for the decade 1949–1959.

By having its sales tied more closely to a market segment whose demand for coal is secularly increasing, the midwestern industry reaps an important benefit: its share of the total market for coal rises as sales to utilities continue to increase. If the demand for coal in all markets other than the utility market remains constant, and sales to utilities continue to grow as they have in the past, the midwest's share of the national coal market will rise from 21.9 percent to 27.8 percent by 1979.

Moreover, relying heavily for sales on a market that is relatively stable reduces the possibility of sharp changes in the price level. As a corollary, this freedom from violent shifts in demand tends to reduce excess capacity, with consequent stabilizing influences on earnings. The stimulus in the eastern coal regions of temporarily heightened demand for export coal can have repercussions that are felt long after the increased demand subsides. Marginal coal operators enter or re-enter the industry in response to an increase in demand, hoping to reap quick profits. Exit from the industry usually proceeds more slowly than entry, and the result is that excess capacity hangs over the industry like a pall.

Midwestern Coal Quality

A determinant of the midwestern industry's ability to compete with rival coal fields is the quality of its coal reserves. Midwestern coals are generally satisfactory for most fuel requirements though they are deficient for some applications.

Coal's most specialized use is in the production of metallurgical coke, an essential element in the manufacture of pig iron. A satisfactory coking coal must contain comparatively low amounts of sulphur and phosphorus to prevent damage to the pigs. Additionally, it should possess the proper agglomeration (coke-making) characteristics. The midwestern region is short on reserves meeting these qualifications, a fact which virtually bars it from one of coal's largest markets. The only commercially minable metallurgical coal in the midwest is located in a small area in southern Illinois, and these deposits are considered marginal. Steel mills in the Chicago area, major metallurgical coal users in the midwestern market, must rely therefore on coking coal shipped in from eastern Kentucky and West Virginia coal fields.

Less specialized are the requirements for steam coals. A wide variety of coal types is useful for this purpose. The best steam coal, though, will be one with a high B.t.u. content and long-burning characteristics. It should contain enough volatile matter to allow a quick response to forced firing. Under the old-style underfeed burning equipment, a high-fusion temperature of ash was required.[1] At the other extreme, the so-called "wet bottom," slag-type furnaces now prevalent in large steam plants demand a low-fusion coal. Spreader stokers and pulverized coal-burning equipment take coals with a larger range of fusion temperature.

The importance of adapting burning equipment to coals with different ash fusibilities may not be readily apparent, but a closer examination reveals an enormously significant trend. Requiring high-fusion temperature coals and ones that caked readily, the underfeed stoker made any given plant captive to eastern mines whose coals possessed these characteristics. In the last two decades, however, innovations in coal-burning equipment, by widening the range of permissible coals, have similarly widened the markets for midwestern coals. Whenever archaic underfeed stoker-fired steam plants located in the midwest's natural market area were replaced with modern spreader and pulverized burner installations, midwestern coal usually replaced the eastern fuel formerly purchased. The expansion of midwestern coal into markets formerly dominated by the east through this means has been pronounced, especially in northern Wisconsin's paper mill region. In 1929 the 1,566,506 tons shipped into Wisconsin from mines in Illinois, Indiana, and western Kentucky compared with shipments totaling 8,243,000 tons from the Appalachian fields. By 1958, the east's share of this market had declined to 6,134,000 tons while midwestern shipments had climbed to 7,080,000 tons.[2]

More exacting standards are erected for tile-burning coals and coals capable of producing gas. Ceramic plants require coal with a low sulphur content since the gases from high sulphur coal can damage the finished ware. Indiana and Illinois have reserves of coal with a sulphur content sufficiently low to meet the demand of plants in this industry within the midwestern market area. Producer gas production also requires a low sulphur coal and in

[1] Coal is usually analyzed on an "as received" and "dry" basis. The "as received" analysis includes the coal's inherent and surface moisture. Occasionally, analyses are determined on a dry, mineral matter-free basis in which the values for the pure coal alone are calculated.

[2] William Harvey Young, *Sources of Coal and Types of Stokers & Burners Used by Electric Public Utility Power Plants* (Washington: The Brookings Institution, Dec. 31, 1930), p. 23, and Bureau of Mines, *Bituminous Coal and Lignite Distribution, 1959,* M.M.S. no. 3035.

addition one with an ash content below 9 percent. Few midwestern seams can meet these specifications, but the demand for this use is still negligible.

The ability to store well is an important factor in coal choice. Coals that weather readily or that are naturally friable create storage problems. The principal changes in stored coal involve either (1) the interaction of coal with oxygen resulting in lower heat values, altered coking characteristics, or spontaneous combustion, or (2) reduction in moisture content causing the coal to break up. Techniques used in storing coal, along with the exposure time and rate of oxygen reaction will determine the extent of these changes. Midwestern coals, with their high moisture contents, suffer the twin perils of spontaneous combustion and crumbling when stored for prolonged periods of time. The result has been some measure of prejudice against these coals in the past by industrial buyers and retail dealers. Retailers generally have refused to purchase midwestern coals during the summer months, even at reduced prices, fearing that the coals would deteriorate in storage. This attitude accentuates the seasonal demand for midwestern coal in this market, and results in idle capacity during off-peak periods. Fortunately for the midwest, improvements in coal-storage techniques have softened the attitudes of industrial purchasing agents toward the storage of midwestern coal. By properly compacting storage piles, interstitial air pockets can be removed and combustion inhibited. Electric utilities, required to accumulate millions of tons of coal, have developed storage techniques to a fine art and now store midwestern coals in vast quantities, encountering no more difficulty with them than they do with higher-rank eastern coals. But to some unsophisticated buyers used to eastern coals there still remain lingering doubts about the storage capabilities of midwestern coals.

Interfuel Competition

The demand for midwestern coal is tempered not only by competition from other coal fields fighting for a share of traditional coal markets but by competition from substitute fuels as well. Several factors determine fuel choice. Cost is a major determinant, but the physical adaptation of a fuel to a particular use is equally important. For example, coal as a base for metallurgical coke is in an invulnerable, competitive position because of its technological superiority in the ironmaking process. The fact that natural gas requires more storage space than other fuels precludes its use in marine transportation or as an automobile fuel. Coal's bulkiness and dirtiness detract from its utility as a domestic fuel. In general, gas is favored because its energy is easily released and because of its low cost in certain industrial applications. Fuel oil releases its heat readily and stores well, but it is handicapped by a relatively high cost.

In the forty-year period from 1918 to 1958 the percentage of total national energy requirements filled by coal fell from 71.5 percent to 27.5 percent. Crude petroleum's share rose from 9.7 percent to 36.3 percent; the natural gas percentage increased from 3.7 to 30.5.[3] Coal lost out relatively and absolutely for two principal reasons. First, oil and natural gas captured huge segments of the market through technological innovation and through changing consumer preference patterns as income levels increased. Thus, development of the diesel locomotive drove coal from the locomotive fuel market, and fuel oil and natural gas replaced coal as a domestic heating fuel. Additionally, coal requirements fell absolutely from improved efficiency in coal consumption. The amount of coal required per ton of pig iron fell from a 1947–1949 average of 2750 pounds to 2012 pounds in 1961. Coal-burning efficiency improved even more spectacularly in the electric utility industry. There, through the use of better combustion and heat transfer techniques, the average number of pounds of coal consumed per kilowatt-hour of electricity produced dropped from 3.20 to .90 between 1919 and 1958.

In those applications in which fuels are substitutable, cost is not always the only factor dictating the fuel choice. The two periods 1935 to 1939 and 1946 to 1950 witnessed a rapid changeover from coal to competitive fuels. During the first of these periods, the average annual price of coal increased one percent and annual consumption fell 3.6 percent. At the same time, distillate fuel prices were increasing at a faster rate—1.6 percent

[3] Bureau of Mines, *Minerals Yearbook, Fuels, 1958,* pp. 4, 5.

yearly—yet the consumption of this fuel rose at an annual rate of 16.4 percent.[4] The pattern repeated itself in the postwar period. Both coal and distillate fuel prices increased sharply but at approximately the same rates; still, coal consumption dropped 3.6 percent annually while fuel oil use increased 12.8 percent yearly.[5]

In the domestic heating fuel market the choice of fuels is often made on the basis of nonprice factors. Cleanliness, convenience, and comfort carry more weight than cost in the householder's fuel choice. This point was made clear in a study of comparative fuel costs conducted by the Bituminous Coal Institute in 1956. In the three cities surveyed which relied on midwestern coal, Chicago, St. Paul, and Memphis, the prices of no. 5 fuel oil were 58, 66, and 49 percent higher, respectively, than retail coal prices; still oil successfully competed with coal.[6] In virtually every city studied outside of the gas-producing regions, fuel oil and gas prices exceeded coal prices, yet the changeover to these competitive fuels continued unabated. The same condition held true on sales by retailers to off-track commercial and industrial users. In from thirty to thirty-eight Illinois and Indiana cities, the unweighted average cost per million B.t.u. to these consumers was 42 cents for coal, 78 cents for natural gas, and $1.02 for no. 2 heating oil.[7] In these markets, too, coal has lost out to substitute fuels despite its price advantage.

The inattention to price in the choice of fuels is less pronounced in industrial purchasing and almost nonexistent in purchases by electric utilities. Table 11 shows the trend in fuel consumption by manufacturing industries from 1929 to 1954 in the six states comprising most of the midwestern market area. Table 12 notes the level of fuel prices in these states for 1947 and 1954, two periods for which we have Census of Manufacturers' data.

The data entering into Tables 11 and 12 have been converted to a common basis using coal equivalents for comparison purposes. The outstanding characteristic of Table 11 is the secular decline in the consumption of coal by manufacturing industries.

[4] Schurr, *Energy in the American Economy, 1850–1975*, p. 80.
[5] Bureau of Mines, *Minerals Yearbook, Fuels, 1958*, p. 127.
[6] Bituminous Coal Institute, *Comparative Fuel Costs*, 1958.
[7] *Ibid.*

TABLE II.   Consumption of fuels by manufacturing industries, selected states, 1929–1954 (thousand tons)

Fuel	Year	Ill.	Ind.	Wis.	Iowa	Mo.	Ky.	Total
Bituminous coal[a]	1954	7,610	4,942	4,036	1,421	1,451	1,650	21,110
	1947	10,338	5,869	4,078	1,778	1,956	1,544	25,563
	1939	10,185	10,997	3,578	1,118	1,686	1,716	30,080
	1929	19,919	16,724	5,541	2,366	3,243	2,008	49,801
Fuel oil in coal	1954	2,550	2,830	780	170	380	120	6,830
equivalent[b]	1947	3,034	2,460	644	127	479	240	6,984
	1939	1,587	1,840	393	55	310	263	4,448
	1929	2,655	1,747	388	176	572	157	5,704
Gas in coal equivalent[c]	1954	11,260	13,100	310	1,304	2,460	706	29,140
	1947	8,480	9,400	138	515	1,061	304	19,898
	1939	4,096	3,446	43	386	743	308	9,022
	1929	1,120	4,216	160	7	346	317	6,166

ᵃ Includes small amount of anthracite in 1954.
ᵇ Fuel oil conversion: 4.2 barrels = 1 ton of coal.
ᶜ Gas conversion: 1 ton of coal equivalent to 50,000 cu. ft. of manufactured gas, 30,000 cu. ft. of mixed gas, and 25,000 cu. ft. of natural gas. Much of this gas is consumed during off-peak demand periods (mostly summer months) at "dump" rates.
Sources: Sixteenth Census of Manufacturers, 1939, vol. I, 349; Census of Manufacturers, 1947, *Fuels and Electric Energy Consumed*, MC203, pp. 134–145; 1954 Census of Manufacturers.

Consumption in this sector fell from 49.8 million tons in 1929 to 21.1 million tons in 1954. Gas consumption, rising from a coal equivalent of 6.1 million to 29.1 million tons in the same period, filled the void left by coal's losses. The consumption of fuel oil in-

TABLE 12.   Fuel costs for manufacturing industries, selected states, 1947 and 1954 (cost per ton equivalent)

Fuel	Year	Ill.	Ind.	Wis.	Iowa	Mo.	Ky.	Weighted average
Bituminous coal	1954	$6.38	$6.34	$9.22	$7.14	$5.85	$4.50	$6.73
	1947	4.86	5.31	8.05	5.39	4.70	5.47	5.62
Fuel oil[a]	1954	14.48	12.69	15.00	14.41	13.95	15.58	13.72
	1947	12.57	11.63	14.22	14.25	10.86	13.08	12.39
Gas[a]	1954	4.30	3.95	21.51	6.57	6.39	9.97	4.73
	1947	2.84	2.48	19.23	5.72	6.26	11.25	3.18

ᵃ Tonnage equivalents computed in Table 11 used to convert oil and gas consumption data into costs per ton equivalent.
Source: 1954 Census of Manufacturers.

creased only moderately, from 5.7 million- to 6.8 million-ton equivalents.

Fluctuations within individual states were not even by any means. The sharpest sales losses to gas competition occurred in Illinois and Indiana where, significantly, the cost advantage for gas over coal was the greatest among the six states studied. The intrusion of gas into coal's Wisconsin markets was minimized, and there coal held a clear price advantage over gas.

Not all of the reduction in coal sales to manufacturing industries represented a total loss to the midwestern coal industry. A great many industrial plants ceased producing their own power and instead purchased power from electric utilities. Table 11 reveals that the consumption of all fuels by manufacturing industries in the six-state area fell from a coal equivalent of 61.7 million to 57.1 million tons, despite an increase in industrial activity during the twenty-five year period under review. Some of the decline resulted from improved fuel-burning efficiency, but most of the change stemmed from the purchase by manufacturing industries of their power requirements. Power sales to manufacturing industries in the six states increased from 5,246 million kw-hrs to 23,245 million kw-hrs between 1929 and 1954.

Table 13 reveals the impact of this development on coal markets in the six midwestern states. Consumption of electricity was

TABLE 13. Consumption of purchased power by manufacturing industries, selected states, in coal tonnage equivalents, 1929 and 1954 (thousand tons)

Year	Illinois	Indiana	Wisconsin	Iowa	Missouri	Kentucky	Total
1954[a]	4,533	2,759	1,531	729	1,175	782	11,509
1929[a]	1,669	852	761	315	545	253	4,351

[a] Electric energy converted to coal equivalent on the basis of 1 kilowatt-hour = 1.66 pounds of coal in 1929 and 0.99 pounds of coal in 1954 (see Bureau of Mines, *Minerals Yearbook, 1958*, p. 127).

Sources: Sixteenth Census of Manufacturers, 1939; 1954 Census of Manufacturers.

converted from kilowatt-hours to tons of coal by applying the appropriate conversion factor for each year, bearing in mind improvements in fuel efficiencies that have taken place over the years. The rapid expansion in the purchase of electric power by industry adds a new dimension to the energy market that is often overlooked. Since in the six states under review coal supplied nearly 90 percent of the fuel consumed in electric generating stations, it stands to gain by this development.

Because fuel costs compose such a large proportion of the total costs of electric power production, price is the primary determinant of a fuel choice. Guided by price considerations, the midwestern electric utilities have overwhelmingly selected coal as their energy source. Tables 14 and 15 indicate trends in the consumption and price of the three fuels competing for shares of the utility markets in the six-state area previously studied. Priced lower than either gas or fuel oil in each of the six states, coal sup-

TABLE 14.   Consumption of fuels by electric utilities, selected states, 1928–1959 (thousand tons)

Fuel	Year	Ill.	Ind.	Wis.	Iowa	Mo.	Ky.	Total
Bituminous coal	1959	17,007	12,504	4,896	1,508	2,511	6,895	45,321
	1954	14,105	6,591	3,395	889	1,413	3,743	30,136
	1947[a]	8,515	4,251	1,721	523	504	683	16,197
	1928	6,121	2,414	849	875	1,201	548	12,008
Fuel oil in coal	1959	0	0	0	0	0	0	0
equivalent	1954	0	0	0	0	11	0	11
	1947[a]	1	0	0	0	27	0	28
	1928	N.A.	N.A.	N.A.	N.A.	N.A.	N.A.	N.A.
Gas in coal	1959	2,262	215	0	1,436	256	132	4,301
equivalent[b]	1954	2,504	170	0	1,171	542	312	4,699
	1947[a]	5	0	0	253	42	0	300
	1928	N.A.	N.A.	N.A.	N.A.	N.A.	N.A.	N.A.

  [a] The data for 1947 appear to err on the low side, although they were secured from the same source as the 1954 and 1959 figures, which are reliable. For these three years the consumption of each fuel was totaled for all plants listed in the F.P.C.'s "Steam Electric Plant Construction Costs and Operating Expenses" report and grouped by states. Apparently the 1947 report failed to include all plants operating in the six-state area. The 1928 data, however, secured from another source are considered reliable. The accuracy of the 1959 figures was verified by cross checking with another source.
  [b] Practically all of this gas is consumed at "dump" rates during the summer months.
  N.A. = Not available.
  Sources: Wm. H. Young, Coal Types in Electric Utilities, pp. 74–77; Federal Power Commission, Steam Electric Plant Construction Costs and Operating Expenses, Annual Reports for 1947, 1954, 1959.

TABLE 15. Fuel costs for electric utilities, selected states, 1947–1959 (cost per ton equivalent)

Fuel	Year	Ill.	Ind.	Wis.	Iowa	Mo.	Ky.	Weighted average
Bituminous coal	1959	$5.46	$5.15	$7.87	$5.66	$5.10	$4.06	$5.42
	1954	5.29	5.09	7.17	5.74	5.35	4.68	5.37
	1947	4.27	4.25	4.09	5.13	4.29	4.64	4.28
Fuel oil	1959	0	0	0	0	0	0	0
	1954	0	0	0	0	9.64	0	9.64
	1947	31.00	0	0	0	7.96	0	8.78
Gas	1959	6.76	6.51	0	6.71	5.51	5.64	6.46
	1954	5.60	7.34	0	6.35	6.25	5.66	5.91
	1947	4.20	0	0	3.96	3.40	0	3.83

Source: Federal Power Commission, *Steam Electric Plant Construction Costs and Annual Operating Expenses*, Annual Reports for 1947, 1954, 1959.

plies 92 percent of the utility fuel requirements in this area. Unlike the situation in the industrial sector, high-priced fuel oil was unattractive to the utilities, and in 1959 it was completely excluded from this market. The threat that gas posed to coal's markets in 1954 appeared to have abated by 1959 as average gas prices continued to rise. Gas distributors have sought, whenever possible, to divert low-priced gas from its use as a boiler fuel to so-called higher uses which command commensurately higher prices.

Elasticity of Demand for Coal

Demand elasticities for coal are difficult to compute statistically though we attempt crudely to derive several of them in this chapter. Notwithstanding the limitation, subjective evaluations can still be brought to bear on the problem.

Elasticities vary in the different contexts in which they are studied. Competing firms within districts and competing districts themselves face kinked demand curves. Small price changes upward result in large tonnage shifts away from the firm raising prices. Below the going price, a firm's demand curve is inelastic, lower price offers being met by competitors anxious to retain their market shares. Similar conditions exist in fringe markets where intradistrict competition occurs. In cases of both intradistrict and interdistrict competition, however, there is a lagged relationship between price changes and shifts in quantities demanded. The possibility of high elasticities above the going price in the short run is damped by contractual relationships between seller and buyer limiting the freedom of each party to act during the life of the contract.

Considered as only one of a family of fuels powering the nation's economy, coal faces an inelastic demand curve. The demand for each fuel, including coal, is largely a derived demand. Changes in fuel prices result in little or no change in the demand for fuel as a source for space heating. In the industrial sector, fuel cost is a minor item in the budgets of all but a few industries. The 1954 Census of Manufactures reveals that the cost of fuel for seventeen two-digit industry classifications ranges from .4 percent to 18.4 percent of the total value added by manufacture for each group.[8] In all but two industries, the figures run less than 5 percent. Changes in fuel prices to most industry groups do not affect industry costs enough to make appreciable shifts in their cost curves. And even if they do, the demand for many industrial products is inelastic also so that realignment of their cost curves would do little to influence industry demand. The demand for fuel *in toto* is much more sensitive to changes in the general level of industrial activity than it is to price-level changes.[9]

We can be somewhat more explicit concerning elasticities for fuel sold to the electric utility market; however, they are derived from the known elasticities of demand for electricity itself and not derived for the fuels directly. A Department of Interior study computes the income elasticity of demand for the Federal Power Commission's Region I which includes New England and the Middle Atlantic States.[10] This is an area which in terms of per

[8] Census of Manufactures, 1954, vol. II, Industry Statistics.

[9] This statement ignores a problem that interests economic historians and students of economic development, i.e., the extent to which the availability of cheap fuel supplies influences a nation's development. It could be argued that instead of the level of industrial activity determining fuel demand, the cost and availability of fuel influence industrial activity. Distinguishing cause from effect in this situation can be difficult. In the present analysis, it is assumed that changes in fuel prices would be comparatively minor, insufficient at least to affect materially the over-all demand for industrial products.

[10] Department of Interior, "Report to the Panel on Civilian Technology on Coal Slurry Pipelines," May 1, 1962.

capita income and industrial activity resembles the market area for midwestern coal. An equation fitted to the logarithms of the variables electricity and real income for the years 1947 to 1960 takes the form Log $y = -1.61387 + 1.89120 \log x$, y being electricity consumption and x, real income. The correlation coefficient is .99663. Thus, a given increase in real income calls forth an 89 percent greater increase in the consumption of electricity. In the absence of technological changes improving the fuel efficiency rate (pounds of coal required to produce a kilowatt-hour of electricity), this increase should create a corresponding increase in the demand for coal to generate the extra electricity.

The same report studies the price elasticity of demand for electricity for the same region. The regression equation for the years 1950 to 1960 is $\log y = .54085 - 1.6357 \log x$, where y is electricity consumption deflated for real income changes and x represents electricity rates. Here, r is .95445. In this case, though, the elasticity of coal demand will not correspond to the price elasticity of demand for electricity. The regression equation indicates that a one percent reduction in electricity rates calls forth a 1.63 percent increase in electricity consumption. But fuel costs account for only 34 percent of total electricity costs.[11] If electricity consumers derive all the benefit from a coal price reduction, then a 10 percent lower delivered coal price results in a 3.4 percent cut in electric rates. These lower rates in turn, generate a 5.54 percent increase in electricity demand. The resulting derived price elasticity for coal demand is −.55. But the *effective* elasticity for the coal operator is even less. Since transportation charges represent approximately 42 percent of total delivered coal costs, a 10 percent lower delivered price on the average spells a 17.5 percent lower mine price. Viewed in this light the relevant price elasticity for the operator faced with the prospect of adjusting price is only −.31.

The most significant measurement of the price elasticity of demand for coal involves its measurement with respect to other fuels. Here, it is useful to analyze each major consumer com-

---

[11] Federal Power Commission, "Statistics of Electric Utilities in the United States, 1961," p. XXII. Unfortunately the 34 percent figure represents fuel costs, not coal costs. It should serve as a fair approximation to the coal-cost figure.

ponent separately, since the over-all cross elasticity is merely a weighted average of the individual elasticities in each of these submarkets.

The demand for metallurgical coal is highly inelastic. Reductions in the cost of blast furnace coke affect pig iron costs inappreciably and certainly not enough to stimulate the demand for iron or the derived demand for coal. On the other hand, since there is no substitute for metallurgical coal in the ironmaking process, increases in its price little influence demand for it. Recognizing its importance in iron production, the major steel companies have protected their positions by acquiring large reserves of metallurgical coal. This practice puts an effective ceiling on the prices of coking coal sold in the commercial market.

In the home heating market coal demand is inelastic in the short run. It is more elastic in the long run when time permits householders to convert burning equipment to take advantage of lower-cost fuels. Cross elasticities, however, are weak even in the long run. Cleanliness, comfort, and dependability more than price determine the average householder's choice of fuel.

The demand for coal in industrial markets is inelastic in the short run since consumers, by and large, are tied to the fuel which can be used in their burning equipment. As equipment wears out, though, changes can be made from one fuel to another; therefore, in the long run, demand becomes elastic. The substitution of one fuel for another is not necessarily made on the basis of minimizing costs. Table 12 reveals that comparatively large quantities of fuel oil are purchased by industrial plants at prices above the average for industrial markets. This phenomenon is due partly to the clouding of individual prices that results from the averaging process. In most cases, however, it stems from the irrational decisions of buyers intent on buying a "modern" fuel.[12] In these instances the responsiveness of demand to coal price changes weakens, and cross elasticities are low.

No such irrationality governs an electric utility's fuel choice.

[12] There are other factors than the delivered price of fuel that enter into the total cost of a fuel. Problems of boiler maintenance, thermal efficiency, labor costs, and so forth, must be considered as well. Even after accounting for these factors, coal costs are generally lower than fuel oil costs at the price levels indicated in Table 12.

With fuel costs accounting for over 60 percent of total production expenses, utilities strive to select the cheapest fuel possible.[13] To provide themselves flexibility in the selection of fuels, most modern generating stations equip themselves with furnaces that can burn any of the three major fuels. Under these circumstances, the demand for coal is highly elastic, limited only by the elasticity of supply of the various fuels and by contractual arrangements that tie a utility to a particular fuel during the term of the contract.

Though most discussions of cross elasticities for fuel in the utility market cite the ability of utilities to convert easily from one fuel to another, no one, to the author's knowledge, has measured the extent to which these interchanges are possible. Table 16 throws light on the issue in the midwestern utility market. The

TABLE 16. Fuel interchangeability in midwestern utilities

State	Coal only		Coal plus other fuel(s)		Total	
	Capacity[a]	Generation[b]	Capacity[a]	Generation[b]	Capacity[a]	Generation[b]
Illinois	1,204	8,540	7,417	30,385	8,621	38,925
Indiana	2,293	14,347	2,884	11,257	5,177	25,604
Wisconsin	1,980	8,778	706	2,138	2,686	10,916
Kentucky	1,731	12,810	740	2,832	2,471	15,642
Iowa	0	0	871	3,611	871	3,611
Missouri	108	94	575	3,727	683	3,821
Total	7,316	44,569	13,193	53,950	20,509	98,519

a Thousands of kilowatt-hours.
b Millions of kilowatt-hours.
N.B. The table covers only utilities consuming either Illinois, Indiana, or western Kentucky coal.
Source: Steam–Electric Plant Factors, 1959, National Coal Association, Washington, D. C., July, 1960.

two breakdowns there total: (1) all plants in the midwest's six-state market area which in 1959 burned only midwestern coal or

[13] We should not confuse this figure with the 34 percent figure previously indicated as the percentage of total electricity costs represented by fuel costs. That total figure includes all costs including production costs, transmission costs, administrative expenses, and depreciation. The 60 percent figure refers only to production (i.e., generation) costs.

were capable of burning only coal, and (2) those plants which in 1959 burned more than one fuel, including midwestern coal, or were capable of burning coal and at least one other fuel. Plants geared to interchange fuels accounted for 64 percent of total installed capacity. Based on electricity generated, the figure fell to 55 percent, a function of the greater utilization of plants designed to burn coal, which is relatively cheaper than its sister fuels in this area of the country.

A word of caution concerning these figures: providing plants with multifuel burning capabilities does not insure that they, in fact, will burn more than one fuel. Reread Table 14. Notice there that in 1959 these plants overwhelmingly selected coal as their fuel choice, coal consumption acounting for 91 percent of their total fuel burn.

It is difficult to measure cross elasticities in the midwestern utility fuel market. In this market midwestern coal competes primarily with natural gas. Utilities purchase much of this gas on an "interruptible" basis, acquiring it during off-peak seasons at dump rates. As temperatures fall and demand for gas in the domestic heating market increases, gas distributors interrupt their service to the utilities and divert gas into the higher-price market. The relative quantities of coal and gas consumed by utilities depend largely, therefore, on the availability of gas, as well as on relative prices.

Along the Atlantic seaboard, interfuel competition centers around coal and residual fuel, imported from the Middle East and Venezuela. Here, oil availability is a problem to some extent, but for the most part oil is rationed through the price mechanism. Residual oil prices fluctuate sharply in response to changes in the world market for oil. This eastern utility market thus provides a better test of demand elasticities than a study of the midwestern market allows. The principles involved, nevertheless, apply equally well in both markets.

Table 17 lists average coal and residual oil prices for the Consolidated Edison Company of New York City, together with the percentage of total fuel consumption represented by coal from 1951 through 1959. A simple linear regression model was used to fit the data (the spread in fuel prices and the percentage of coal

TABLE 17. Fuel prices paid and coal consumed by Consolidated Edison Company, 1951–1959

Year	Avg. coal price[a]	Avg. fuel oil price[a]	Price difference (col. 3 − col. 2)	Coal consumption (percent)
1951	35.1	37.0	1.9	79.9
1952	35.8	36.8	1.0	78.2
1953	36.9	33.6	−3.3	70.1
1954	39.2	34.8	−4.4	68.7
1955	35.3	37.8	2.5	80.5
1956	37.7	43.1	5.4	82.1
1957	39.2	48.8	9.6	80.5
1958	39.5	39.2	−0.3	76.4
1959	37.3	36.3	−1.0	73.5

[a] Prices figured in cents per million B.t.u.

Source: Federal Power Commission, *Steam Electric Plant Construction Costs and Annual Production Expenses,* Annual Supplements 1951–1959.

consumed) for these nine years by the least-squares method. This regression had a correlation coefficient of .86, statistically significant at the one percent level. By excluding the data for 1957, a year in which supplies and prices were distorted in the backwash of the Suez crisis, the correlation coefficient increased to .974.

Although the regression gives a close fit to the data, a word of caution is required. If the elasticity of demand were sensitively related to price fluctuations, shifts away from one fuel to another would be complete when one gained a price advantage over the other. For example, in 1952 coal should have commanded 100 percent of the market, but it should have lost *all* its share the following year when residual oil prices fell below the average coal price. If shifts of this order of magnitude had occurred, cross elasticities would have been high, yet they would have been masked if the relationships had been tested by means of a correlation coefficient as was done above. *All* positive differences in price in favor of coal would have been associated with a 100 percent use of coal, negative values, with zero use.

That elasticity is not infinitely elastic results from two factors. First is the previously stated condition that suppliers and consumers are bound by contractual obligations that limit their short-run freedom to respond to price changes. In addition, there are inflexibilities in both coal and residual oil supply schedules.

As previously indicated, a measure of historical demand elasticities in the midwestern utility market cannot easily be made. By making some simplifying assumptions, however, hypothetical elasticities can be constructed to shed light on the responsiveness of coal demand in this market to changes in the price of coal.

Assume that the price of gas purchased by utilities remains unchanged as coal prices are altered. Assume further that the midwestern coal industry practices price discrimination by reducing prices 10 percent to those utilities consuming gas at costs below the delivered-coal costs, with coal prices to other utilities remaining unchanged. Coal-purchase decisions are assumed to be made solely on rational cost considerations. Within this framework, what effect would changes in coal prices have on the demand by midwestern utilities for midwestern coal?

The same six-state area referred to in Table 15 was studied for 1959. A selective 10 percent price reduction would have increased sales from 45,321,000 tons to 46,684,000 tons, an increase of 3.02 percent. The average delivered price on all utility sales would have fallen from $5.42 a ton to $5.37, a reduction of 0.92 percent. Price elasticity would have been −3.26.

An additional 10 percent price reduction to other utilities consuming gas (keeping the first price reduction at 10 percent) would have given an over-all tonnage increase of 3,315,000 tons. The average delivered price on utility sales would have fallen to $5.22; elasticity would have been −1.98. (In both cases we have ignored in the elasticity computations the derived demand effect on coal consumption stemming from a reduction in electricity costs; instead, we have focused solely on the substitution effect of coal price reductions vis-à-vis oil and gas.)

Within this framework, then, demand is elastic just below existing prices, becoming less elastic as prices are further reduced. At some point the demand curve for coal becomes infinitely inelastic when coal has captured all of the market.

A somewhat different pattern of elasticity exists for the upper portion of the demand curve. Moderate price increases result in only moderate tonnage losses. Gas unavailability and noncompetitive fuel oil prices combine to make this part of the demand schedule inelastic. At still higher prices, however, the curve becomes elastic again as the impact from fuel oil and nuclear power competition is felt.

What general picture of elasticity emerges from the foregoing analysis? The demand for coal being largely a derived demand, changes in coal prices little affect the demand for coal. The extent to which this relative inelasticity prevails depends on the markets under consideration, demand being less price-inelastic in utility markets than in manufacturing industries where coal costs represent a smaller percentage of total production costs. There is evidence that coal demand is relatively elastic with respect to income in the utility market and doubtless in general industrial markets as well. Cross elasticities are moderately high, modified, however, by some degree of gas unavailability and by fuel contract arrangements that distort the picture.

Summary of Competitive Forces

Chapters II and III have studied the relevant factors affecting the demand for midwestern coal, along with some of the forces in and out of the industry modifying market dominance in its market area. This section briefly summarizes the nature and extent of these competitive forces.

Any midwestern coal producer faces potential competition on three fronts: he competes with producers of alternative fuels, with coal producers from outside coal regions, and with coal operators within his own region. Chapter II stresses the forces limiting interregional competition in the midwestern industry's base market. Over 75 percent of the midwestern production moves into markets virtually free from outside coal competition, competition-free markets being defined as those in which the midwest furnishes more than 95 percent of the area's coal requirements. In some parts of the protected market area the midwest sufficiently dominates the coal scene to warrant labeling the areas as regional monopolies.

This chapter explains the extent to which price competition prevails in markets in which alternative fuels can compete with coal.

Nonprice factors largely determine fuel selection in the household market in the long run; in the short run demand is relatively inelastic. Electric utilities are sensitive to price differences in the short run as well as the long run, but their ability to move away from coal is limited by the inelasticity in the supply of gas. Fuel oil prices to this consumer class are too high to offer coal serious competition. Price competition in the industrial field prevails, but here again gas supply availability is a limiting factor. Thus, interfuel competition is far from complete. It would be a mistake to assert that midwestern coal operators need not consider the effect of their price decisions on competition with alternative fuels; nevertheless, the market factors cited above appreciably diminish the importance of interfuel competition.

Interregional and interfuel competition being severely limited, there remains principally intraregional competition to check the midwestern industry's market power. (Indeed, in the case of any coal-producing region, the presence of *any one* of the three factors is sufficient to limit market power.) Chapter IV begins a study of the factors affecting intraregional competition, focusing attention first on the concentration of production in the midwest. We shall determine whether the forces insuring healthy competition within the midwest have been weakened in any way. If limited intraregional competition joins weak interregional and interfuel competition, the ring will have been closed, and the midwestern coal industry will have fashioned for itself a powerful structure for dominating the midwestern fuel market.

CHAPTER IV

The Structure of the Midwestern Coal Industry

Until now, we have viewed the market from two standpoints: as a group of consumer classes (i.e., electric utilities, retailers, etc.) and as a geographical area. We have also marked off the midwestern coal market from substitute fuels, defining the limits of interfuel competition. These conceptions deal with two of the three minimum features required of a market in theory and in practice: the existence of buyers and a point, line, or plane where transactions occur. But they neglect the seller, a key participant in any market transaction. This chapter remedies this deficiency by studying in detail the composition of sellers in the midwestern coal industry.

Additional characteristics of the midwestern coal market structure will be dealt with as well. In Chapter I, market structure was defined as the organizational factors which influence pricing and competition within a market. Already we have studied the dimensions and relative isolation of the midwestern market area. This chapter carries the analysis of structural characteristics several steps further by studying the degree of seller concentration and the size distribution of sellers of midwestern coal and, as a corollary, the methods used in achieving the present degree of seller concentration; the extent of vertical integration among midwestern producers; and the size distribution of buyers in the major consumer classes. Reserved for Chapter VI is a study of further structural characteristics including conditions of entry into and exit out of the industry and the pattern of coal reserves ownership; the degree and value of product differentiation in the industry; and the nature and influence of prevailing contractual arrangements between buyers and sellers. We study first the degree of seller concentration and size distribution of sellers in the midwestern industry.

Industry structures fall within the limits of the two extreme market forms, pure competition and monopoly. At the one extreme, the actions of any firm have no appreciable effect on the price-output decisions of other firms in the industry. Each firm in Tibor Scitovsky's parlance is a price-taker. At the other pole, the monopolist, being the sole producer of a commodity, is free to make whatever price-output decision will maximize his profits. Between the extremes are market forms possessing various degrees of market imperfection, in which the extent of market power (measured in terms of control over supply) varies with a number of factors, including the degree of seller concentration. As the number of firms in the industry increases, presumably the actions of each directly affects his competitors to a diminishing extent. Under oligopoly, however, each firm controls a sufficiently large share of the market to make his price-output decisions felt by all others in the industry. An attempt should be made, therefore, to study the numbers and sizes of firms in the industry to determine the market category into which the industry falls and to test the relevance of accepted theory in predicting industry behavior.

Measures of Concentration

Clearly there are several ways to measure concentration. Numbers of employees, total assets, sales or physical output, and value added by manufacture offer four possible measuring rods.[1] An analysis of market power manifestly calls for concentration ratios based on sales or their equivalent. In a study of the midwestern coal industry we need a substitute for dollar sales figures. With less than half of the midwestern industry's tonnage being produced by companies with listed securities, data on dollar sales

---

[1] See M. A. Adelman, "The Measurement of Industrial Concentration," in *Readings in Industrial Organization and Public Policy,* ed. George W. Stocking and Richard B. Heflebower (Homewood, Ill.: Richard D. Irwin, Inc., 1958), p. 8.

are incomplete.[2] But coal production, expressed in tons, serves effectively as a proxy variable. It has the advantage of being readily available in industry statistics and of being accepted in the trade as the leading indicator of concentration.

Substituting production figures for sales totals creates little distortion in the concentration picture. What little distortion there is probably takes the form of lower ratios when computed on the basis of production rather than on sales. The larger mines are more apt to upgrade their raw coal by washing than are their smaller competitors, leading to above-average realizations for the top firms. On the other hand some of the smaller firms are forced by high costs to seek out markets where average prices are above normal, pulling their realizations above the industry average. On balance, though, the larger (and generally more prosperous) firms probably enjoy slightly higher average realizations than the smaller firms, depressing concentration ratios based on tonnage somewhat below those based on sales. The realization figures necessary to support this view being unavailable, we caution that the foregoing view is an impressionistic view, albeit a realistic one.

Usually concentration ratios, expressed as percentages of output produced by one firm or groups of firms, are based on national output. This convention can be misleading by ignoring the market protection afforded by the inability of distant firms in industries bearing high transportation charges to compete effectively in nearby markets. Chapter II described the regional character of the midwestern market and justified measuring concentration only among the midwestern firms supplying this regional market.

Concentration of Coal Production

We need to study two aspects of concentration. Our essential task is to measure concentration ratios. Additionally, we must look briefly at changes in the concentration of output among the

---

[2] In 1959 Moody's *Industrial Manual* listed financial data on the following companies: Peabody Coal, Truax–Traer Coal, West Kentucky Coal, Ayrshire Collieries, Old Ben Coal, Bell and Zoller Coal, and United Electric Coal. Together these firms produced 49.6 millions of tons of coal from their midwestern operations, or 55.3 percent of total Illinois, Indiana, and western Kentucky production.

various mine size groups. The mine is, after all, the basic economic unit in the industry.

There is a decided trend toward concentration of production among larger mines in the midwest. In 1959 the proportion of total output produced by large Illinois, Indiana, and western Kentucky mines (with annual outputs exceeding 500,000 tons) ranged from 62.9 to 87.6 percent. Western Kentucky has shown the most dramatic shift in emphasis toward larger-scale operations. There the proportion of total output coming from mines producing over 500,000 tons annually increased from zero in 1934 to 74 percent in 1959 as new large mines were developed to supply the burgeoning electric utility market. As production shifted toward larger mines, it moved away from smaller operations. Again, in 1959 the midwestern mines producing less than 50,000 tons annually accounted for only 1.8 to 3.8 percent of their state's production, a decline from the 7.9 to 9.1 percent figures achieved by the smaller mines twenty-five years earlier.

The effect of this greater concentration of midwestern production on the number of mines in the industry is shown in Table 18

TABLE 18.   Number of midwestern coal mines, 1944–1959

Year	Indiana	Illinois	W. Kentucky	Midwest
1959	80	137	115	332
1954	98	185	126	409
1949	117	299	308	724
1944	152	325	370	847

Source: Bureau of Mines, *Minerals Yearbook*, various years.

which indicates the number of midwestern mines producing over 1,000 tons yearly in the period 1944–1959. Though midwestern production declined 28 percent between 1944 and 1959, the number of midwestern mines fell 61 percent. Concentration of production in a few large mines intensified between 1949 and 1959 when the number of midwestern mines declined 54 percent despite a 10 percent increase in midwestern production.

This trend in the midwestern industry's structure of mine size is bound to have repercussions on its performance. First, it should affect entry considerations by influencing the financial requirements necessary for entering the midwestern industry. Furthermore, if economies of scale prevail in the industry, midwestern operators who expand their output should move downward along their long-run cost curves. We pursue these matters in detail in succeeding chapters.

Useful as concentration data figured on a mine basis may be, they are deficient in an important respect. They fail to show the concentration of economic power controlled by decision-making units. Indeed, this deficiency exists in nearly all published studies of the coal industry dealing with the concentration problem. It results partly from the character of government statistics that use the mine, not the company, as the reporting unit. Even if this defect were remedied a further problem would exist. Many mine operators produce coal under different corporate names. Often, in the premining stage of development, virgin coal land is held by a nonoperating, landholding subsidiary of a parent mining company. For tax or other reasons, the new corporation continues to exist as an operating subsidiary after mining operations begin. Sometimes subsidiary companies are created to provide key employees an opportunity to reap capital gains tax benefits. Capitalization of the landholding subsidiary can be kept low, with key employees offered inexpensive shares of the firm's common stock. Loans from the parent corporation to the subsidiary (or institutional loans guaranteed by the parent) provide necessary funds to commence mining operations. The resulting leverage on the common stock greatly enhances its value if the operation is successful.

Corporate entities thus can proliferate, deceiving those concerned with measuring concentration in the hands of decision-making groups. Failure to recognize the distinction between *companies* and decision-making *groups* (i.e., companies under common ownership or control) led one student of the industry into serious error. Hubert F. Risser's study, *The Economics of the Coal Industry* is based on the fallacious notion that the "company" can be used as the unit for measuring industry concentration.[3] As a result his conclusions are questionable.

This study of the industry corrects previous deficiencies by making the company-*group* rather than the company itself the focal point of interest. In the case of most smaller firms the concepts were congruent since these firms usually operated a single mine under one company name. Larger firms controlling one or more subsidiaries were grouped into single units headed by the parent company responsible for making the firms' basic policy decisions.

Following the Census convention, Table 19 shows the concen-

TABLE 19.   Concentration of midwestern coal production among top four, eight, and twenty company-groups, 1934–1962 (tonnages in thousands)

Year	Four largest		Eight largest		Twenty largest	
	Tons	Percent	Tons	Percent	Tons	Percent
1962	52,169	54.6	70,851	74.2	N.A.	N.A.
1960	48,833	52.3	65,122	69.7	83,278	89.2
1954	19,687	25.2	32,224	41.4	54,580	70.0
1949	21,908	26.8	32,115	39.3	50,217	61.4
1944	31,254	25.2	45,031	36.3	71,861	57.9
1938	14,454	22.6	21,662	33.8	35,739	55.9
1934	12,680	19.7	19,500	30.4	32,450	50.5

N.A. = Not available.
N.B. The figures exclude small amounts of captively controlled tonnage.
Sources: Coal Production in the United States, compiled by Keystone Coal Buyer's Manual (New York: McGraw-Hill Publishing Co., Inc., 1944–1960); Annual Reports, Illinois, Indiana, and Kentucky Departments of Mines and Minerals, 1934 and 1938; Indiana Coal Trade Association, *Coal Production in Indiana, 1926–50*; Midwest Coal Producers Institute, Inc., *Report of Mine Performance, 1962.*

tration of output among the top four, eight, and twenty company-groups in the midwestern industry. The trend toward concentration is unmistakable. In 1934, the top twenty company-groups

[3] Lawrence, Kansas: Bureau of Business Research, University of Kansas, 1958.

controlled 50.5 percent of midwestern production; by 1960, this figure had risen to 89.2 percent. Control by the four leading groups had increased correspondingly. Controlling 19.7 percent of midwestern production in 1934, the four largest company-groups accounted for 54.6 percent of midwestern tonnage in 1962. Note that most of the increase in concentration occurred in the most recent years, from 1954 to 1962.

The top eight midwestern producers share unequally in their control of the market, as indicated by Table 20. Peabody Coal

TABLE 20. Distribution of production among eight leading midwestern coal company-groups, 1962 (tonnages in thousands)

Company-group	Tonnage[a]	Percent of midwestern production
Peabody	26,548	27.8
Freeman–United Electric[b]	11,689	12.2
Ayrshire	7,958	8.3
Midland Electric	5,974	6.3
W. Kentucky	5,568	5.8
Consolidation (Truax–Traer Div.)	5,021	5.3
Bell and Zoller	4,289	4.5
Old Ben	3,804	4.0
Total	70,851	74.2

[a] Tonnage figures refer only to production of the company groups from their Illinois, Indiana, and western Kentucky mines.

[b] United Electric Coal Co. and Freeman Coal Co. are lumped together as a result of Freeman's control of United Electric.

Source: Midwest Coal Producers Institute, Inc., Report of Mine Performance, 1962.

Company controls over 27 percent of the midwestern production, and Peabody and Freeman–United Electric together account for two fifths of midwestern output. The comparative dominance by one firm is a fairly recent phenomenon in the midwest. As late as 1954, the largest operator shipped only 8.2 percent of the midwestern output.

Concentration can also be measured by the number of firms accounting for a specified percentage of industry output. Using this measuring device we still find control within the midwest becoming more concentrated. In 1944 fifteen company-groups controlled 50 percent of the midwestern output. By 1962 the number had been reduced to four. Here again, most of the concentration has occurred since 1954, when there were still eleven firms controlling half the midwestern production.

Nothing in the structure of the channels of distribution vitiates the growing market power flowing from the increasing concentration of production in fewer hands. A survey conducted by the author among midwestern coal operators throws light on the issue. Ninety-five percent of the production controlled by the top twenty firms came from company-groups producing over one million tons in 1960. Company-groups accounting for 83 percent of this production responded to the survey's questions relating to distribution channels. These operating groups sold 91.8 percent of their production directly, using their own sales forces or wholly owned subsidiary sales agencies. The remaining tonnage moved through independent sales companies and brokers. In these cases, control over prices usually remained unfettered in the hands of the producer. In some instances the producer established minimum prices, the independent sales agency being empowered to sell coal at any price it saw fit above the minimum. Thus, there is no evidence that the structure of marketing channels weakens market power in any way. Indeed, it may strengthen it. Some of the top twenty producers share common sales agencies. Though, as we have shown, control over prices usually rests with the producer, the sharing of common sales facilities raises unique opportunities for these firms to coordinate their sales approaches.

The emphasis throughout this discussion has centered on the tightening of market power resulting from the increasing concentration of production in fewer hands. The net effect of this trend has been to create an oligopolistic market structure composed of medium- to large-sized firms, each sufficiently large to make its market adjustments felt by its competitors. Preoccupation with concentration tendencies, however, should not divert our attention from the potential importance of the remaining small firms. Surrounding the oligopolistic "core" is a "competitive fringe" of several hundred small firms accounting for 10 to 15 percent of midwestern production. Though their numbers are rapidly declining, they nonetheless pose a potential threat to the security of large producers. Their market position deteriorating, these small operators can be expected to rebel against efforts of the industry to coordinate price and output decisions, and this rebellion should make such coordination generally less effective in fixing prices and quantities.

The increased concentration of production described here resulted from the growth of the larger firms. Some of this expansion was internally generated, but much of it stemmed from external growth. That is, firms grew by joining the assets of two or more firms through mergers or acquisitions. Before studying the timing and extent of external expansion in the midwest, we need to look into the motivations for mergers generally. In so doing we may shed light on their extensive formation in the midwestern industry in recent years. In the following discussion, the word "merger" applies to the joining together of the assets of two or more firms, irrespective of the legal means in fact used to accomplish it.

Motives for Merger

Initiative for a merger may originate either with the buyer or the seller. Several considerations can influence the decisions of either party. Among those affecting sellers are management or personal factors, investment factors, and tax considerations.

Numerous personal and management factors influence decisions to merge. The owner-manager of a small firm may wish to retire. A merger allows him to go into retirement and recover his investment at the same time. Often management is weakened by the loss of key personnel, and a merger provides the only means of protecting the owner's investment. In other cases mergers furnish the best solution to a situation in which dissension among top personnel has seriously weakened management. Sometimes an owner may simply wish to be connected with larger companies, as a security measure, or because he feels the opportunities are greater. Finally, failure to keep pace with competition may lead to a merger as a means of salvaging the firm's assets.

Several investment considerations influence sellers' merger decisions. Sellers may take the initiative because they are less optimistic about the industry's future than are prospective buyers. Recognizing the probability that the value of the firm's assets may decline, the owner may prefer to merge. Doubtless this factor has influenced some coal mergers. Occasionally an owner-manager may wish to earn a quick profit by capitalizing the firm's earnings and securing a price for the firm that exceeds its net worth. In some instances the motive is a desire for diversification.

Tax considerations loom large in the decisions of sellers to merge or sell out. Often a merger solves an inheritance tax problem by converting an unmarketable security into cash or into the security of a listed corporation. At the worst the seller pays only a capital gains tax on any profits accruing to him from the sale. If an exchange of stock seals the merger, the transaction is tax-exempt to the seller.

For our purposes, it is perhaps more important to analyze motives of buyers who are responsible for mergers than it is to investigate sellers' motives. In their study of mergers, J. Keith Butters, John Lintner, and William L. Cary report on the aims of several buyers involved in mergers. Some were interested in acquiring new products and new plant capacity. Others sought through mergers to achieve self-sufficiency by means of vertical integration. In still other cases, buyers wanted to improve the marketability of their stock to make outside capital more readily available. Some were simply "buying" working capital.[4]

Obviously there are additional merger motives. Edith Penrose writes:

The fact that fixed capital, market position or even goodwill, once created, persist in time means there is always a choice between recombining the old and building anew. If legal institutions permit one going concern to purchase or merge with another, then whether expansion takes

[4] J. Keith Butters, John Lintner, William L. Cary, *Effects of Taxation on Corporate Mergers* (Boston: Harvard Univ., Graduate School of Business Administration, 1951), pp. 214–222.

place through the building of new plant or through the acquisition of another firm will, in any particular case, depend on which appears to be the more profitable course of action. Whenever merger is considered the more profitable way to expand, there will surely be a tendency for merger to occur.[5]

Penrose's first alternative — building new plant — adds to industry capacity. The acquisition solution, by shifting ownership of existing facilities from one producer to another, reduces competition by eliminating a competing seller from the market. Thus there may be a link between the added control over supply provided by the merger and the "more profitable" outcome Penrose refers to. In fact, the merger may manifest a conscious policy to improve profit margins through enhanced market control. On the other hand, the acquiring firm may recognize the advantages of increasing profits through mergers, apart from those flowing from tighter control over supply. For example, the buyer may view merger as a way to acquire new facilities at less than reproduction cost. This alternative may be superior to constructing new plant if the acquiring firm can rejuvenate a "sick" company lacking only competent management to achieve a satisfactory profit rate.

Furthermore, by buying its way into a market, the surviving firm in a merger reduces risks by acquiring a product which presumably has already been successfully tested in the market. And if the growing firm seeks to expand rapidly, a merger accomplishes the purpose faster than building new plant would provide.

An important determinant of mergers is the desire to achieve economies of production. Penrose divides these into technological and managerial economies.[6] Technological economies may result from better specialization of labor, the use of automated equipment, or the installation of larger machines which increase volume and cut costs. Managerial economies result "when a larger firm can take advantage of an increased division of managerial labor and of the closely allied mechanization of certain administrative processes; makes more intensive use of existing managerial resources by the 'spreading' of overheads; obtains

[5] Edith Penrose, *The Theory of the Growth of the Firm* (Oxford: Basil Blackwell, 1959), p. 155.
[6] Penrose, pp. 90–92.

economies from buying and selling on a larger scale; uses reserves more economically; acquires capital on cheaper terms; and supports large scale research." [7] Often expansion takes the form of adding additional plants, similar to those the firm already operates. George Stigler questions the ability of firms to achieve economies from the resulting multiplant operations. "What possible efficiency," he asks, "comes from having a steel plant in Pennsylvania and one in Alabama and several out west? . . . Of course you need only one legal counsel instead of several, and you can perhaps have one central research agency instead of several, but the functions of this type are not quantitatively important." [8] J. F. Weston disagrees.[9] He cites the possibility of large-scale firms reducing costs by eliminating duplicative distribution facilities. And he summons Joseph Schumpeter to defend the thesis that technical advances from research and development expenditures can have an important bearing on the firm's total costs.

Several more important buyer motives for merger exist. The desire to grow because of the prestige attached to size cannot be overlooked, nor can the desire for market control that results from the elimination of competitors. In an extractive industry like coal mining, market control can become even greater if mergers aim at resource domination. Acquiring a competitor with attractive coal reserves reduces competition in the short run, and strengthens the long-run monopoly power of remaining firms in the industry by reducing the threat of entry.

Midwestern Coal Mergers

We need now to analyze recent midwestern coal mergers to determine (1) their timing; (2) the relative importance of external and internal expansion in the growth of concentration of midwestern coal production; (3) the relation, if any, between

---

[7] *Ibid.*, p. 92.

[8] Testimony of George Stigler in *Study of Monopoly Power, Steel.* Hearings before the Subcommittee on the Study of Monopoly Power of the Committee on the Judiciary, U. S., 81st Congress, 2nd Session, 1950, p. 123.

[9] J. Fred Weston, *The Role of Mergers in the Growth of Large Firms* (Berkeley: Univ. of Calif. Press, 1953), p. 70.

merger activity and stock market price levels; and (4) the buyer and seller motivations prompting mergers.

Table 21 summarizes midwestern merger activity from 1950 through 1962. This list, it is believed, includes all of the major mergers and acquisitions entered into during the period under review.[10] The list grew out of a search of state mine reports, coal association tonnage records, and Moody's *Industrial Manuals,* and was double checked by interviewing top executives of many of the leading midwestern operators. It unquestionably includes all midwestern mergers among the members of the Midwest Coal Producers Institute, Inc., a trade association representing approximately 90 percent of midwestern production. If there are any exclusions among the remaining firms they cannot involve a great deal of tonnage, being limited mostly to small truck operations producing less than 50,000 tons annually. It is difficult to trace acquisitions by firms in this size class. Traditionally, however, mines of this size have accounted for little merger activity.

External growth obviously accounted for most of the increase in concentration ratios in the midwest.[11] Seven of the nine companies listed in Table 21 still surviving at the end of 1962 were among the eight largest midwestern producers in 1962. Their combined midwestern production in the years immediately prior to their first mergers was 23,198,000 tons. In 1962, they produced 68,851,781 tons from their midwestern mines. All but 9,103,000 of the more than 45,000,000-ton gain achieved by these companies resulted from external growth.[12]

Precise as this division between external and internal growth may appear, it nevertheless requires a qualification. The data do not permit the type of analysis that would make the division between external and internal growth unassailable. In some cases

[10] We have excluded the merger of Pittsburgh and Midway Coal Co. into Spencer Chemical Co. since this was an internal merger between two firms commonly controlled.

[11] External growth is defined as growth resulting from mergers.

[12] The figures exclude 2,855,000 tons involved in the Northern Illinois Coal Corp. merger with Peabody in 1950. Northern Illinois actually merged with the Sinclair group which subsequently merged with Peabody. It would be erroneous to consider that this tonnage represented growth to Peabody when Peabody and Sinclair did not merge until five years after the Northern Illinois–Sinclair merger.

TABLE 21. Summary of midwestern coal mergers, 1950–1962

Acquiring company	Company acquired	Date of acquisition	Acquired company's midwestern tonnage (year before merger, thous. tons)	How acquired
Peabody	Northern Ill.	Dec., 1950	2,855	Cash purchase of company
	Sinclair Mines[a]	July, 1955	6,042	Exchange of stock
	Wasson	Dec., 1955	250 (est.)	Cash purchase of certain assets
	Midwest–Radiant	Mar., 1956	1,578	Cash purchase of company stock
	Perry Coal	Mar., 1956		
	Energy Mine (Morgan)	Sept., 1956	411	Cash purchase of mine
	Poplar Ridge	Mar., 1957	794	Cash purchase of company stock
	Terteling Bros.	May, 1957	1,020	Cash + exchange of stock
	W. G. Duncan	Feb., 1958	840	Exchange of stock[b]
	Maumee	Jan., 1959	1,655	Purchase of net assets
	Forsythe–Carterville	1961	223	Unknown[c]
	Hart & Hart	Sept., 1962[d]	511	Cash purchase of net assets
Truax–Traer	Binkley	June, 1950	1,845	Cash and exchange of stock
	Pyramid	June, 1950		
	Little Sister	May, 1956	662	Exchange of stock
Consolidation	Truax–Traer	April, 1962	4,472	Exchange of stock
Ayrshire	Minnehaha Mine (Little Betty Mining)	1952[d]	325	Cash purchase of mine
	Carmac	Mar., 1956	297	Cash acquisition of company
	Friar Tuck Mine (Sherwood–Templeton)	Jan., 1960	401	Cash purchase of mine

Bell & Zoller	Consolidated Bradbury Mine (Mid-West Utilities)	Mar., 1951 Jan., 1957	1,711 847	Purchased 4 mines for cash Cash purchase of mine
Freeman Coal Mining	Chicago, Wilmington and Franklin United Electric	Dec., 1954 Nov., 1959[e]	3,222 3,530	Purchased 90% of stock Acquired working control of stock through open market purchases
Enos Coal Mining	Blackfoot Mining	Jan., 1961	596	Cash purchase of company stock
West Kentucky	Nashville	Sept., 1955	3,695	Cash purchase of assets including royalty agreement on coal land leased
Sherwood–Templeton	Pioneer	April, 1961	140	Cash purchase of mine
Snow Hill	Mid-Continent	1959[d]	900[f]	Cash purchase of mine
Midland Electric	Viking, Snow Hill, Mid-Continent, Bledsoe	May, 1962	2,281[g]	Exchange of stock
Total tonnage			41,103	

[a] Includes Sentry Royalty Co., Power Coal Co., Homestead Coal Co., Sinclair Coal Co., Key Coal Co., Broken Aro Coal Co., Alston Coal Co. and Rogers County Coal Co., all controlled by the Sinclair Coal Co. interests. The above companies merged into Peabody Coal Co., but through the exchange of stock the Sinclair interests assumed control of the surviving company.

[b] Duncan received Peabody stock in exchange for half the Duncan stock; the remaining stock exchanged for a 50 percent interest in River Queen Coal Co. and a 20 percent interest in Peabody–Southern Coal Co., both subsidiaries of Peabody. Duncan later exchanged their interests in the subsidiary companies for more Peabody stock.

[c] Purchase arrangement unknown. The acquisition announcement declared that Peabody and Forsythe–Carterville would operate under a "joint agreement." The mine was immediately closed down, output shifting to Peabody's Energy mine adjacent to Forsythe–Carterville's property.

[d] Month of merger unavailable.

[e] This is the date when Freeman presumably gained control of United Electric.

[f] Estimate.

[g] Includes 822,111 tons from Mid-Continent which previously merged with Snow Hill Coal Co.

Sources: Moody's Industrial Manuals, various issues; private correspondence; personal interviews.

the surviving corporations have shut down mines shortly after acquiring them by purchase or merger. The data, therefore, understate the companies' subsequent internal *physical* growth. Presumably, however, the surviving companies acquire the sales contracts of the merged companies along with their physical assets, so that using our definition of concentration based on tonnage *sold,* the merged tonnages still contribute to external growth.

An additional factor modifies the distinction between external and internal growth. In addition to acquiring mining equipment and sales contracts, the surviving company often gains valuable coal reserves as well. Tonnage gains made subsequent to a merger may be internally created, but often they stem directly from the external growth that furnished the coal reserves necessary for expansion. Thus, much of Peabody's internal growth flowed from the Sinclair–Northern Illinois Coal Corporation merger in 1950 that gave Sinclair (later Peabody) an expanded coal reserve base. Similarly, Peabody's western Kentucky tonnage increased far beyond the 840,000 tons of annual production acquired from Duncan Coal Company as a result of the subsequent mine development of idle Duncan reserves included in the merger.

Other cases raise additional measurement problems. As previously noted, acquired mines are often shut down but new mines may be opened on reserves acquired by merger. Does the new mine production represent internal or external growth? Would the answer differ if the new mine used coal reserves not associated with a merger? The problems are numerous and prevent a precise differentiation between external and internal growth. Nevertheless, despite the measurement problems, we can conclude that most of the increased concentration has stemmed directly or *indirectly* from merger activity.

In their study of mergers, Butters, Lintner, and Cary found buyers seeking out sellers one sixth of the time with sellers taking the initiative in almost two thirds of the cases.[13] This being true, the level of stock market prices may influence merger activity since high stock prices give potential sellers a chance to dispose of their holdings at inflated values.

The hypothesis that merger activity and stock prices are cor-

[13] Butters, Lintner, and Cary, p. 309.

related has been tested in previous general studies of merger movements. The correlation between these phenomena for the period 1919–1941 was statistically significant at the one percent level; the correlation coefficient equalled .676.[14] Several bull markets have coincided with waves of merger activity. Two of the most pronounced coincidental upward swings in stock prices and merger activity occurred in the booms of 1897–1899 and 1926–1929. Both periods witnessed the active intervention of the professional promoter who stood to gain from the mergers he created. The argument runs that high stock prices (and presumably the lure of even higher prices) greased the skids on which the mergers were launched.

But the evidence of a close relation between stock prices and merger activity is inconclusive. We find periods when the two did not move together. For example, relatively stable stock prices accompanied the minor merger movement during the period 1940–1947. And probably another way of saying the same thing — the foregoing correlation coefficient is not impressively high. Moreover, forces coinciding with changes in stock prices may be as responsible as the stock price level itself in stimulating merger activity.

With this background we test the hypothesis that coal merger activity varied directly with stock market prices. Figure 7 traces the relationship between the phenomena in the midwest from 1949 through 1962, plotting the number of mergers against an index of midwestern coal stock prices.

The acquisitions of Perry Coal Company and Midwest–Radiant Coal Company by Peabody in 1956 and of Binkley Coal Company and Pyramid Coal Company by Truax–Traer in 1950 are each considered to be one merger. In both cases, although two separate companies were involved on the selling side, control rested with single financial interests, making the transactions single mergers for our purposes. The coal stock price index was constructed by first securing the average low and high prices for Peabody, United Electric, Truax–Traer, and West Kentucky Coal Companies for

---

[14] Quoted in chapter by Jesse W. Markham "Survey of the Evidence and Findings on Mergers," in *Business Concentration and Public Policy* (Princeton: Princeton Univ. Press, 1955), p. 153.

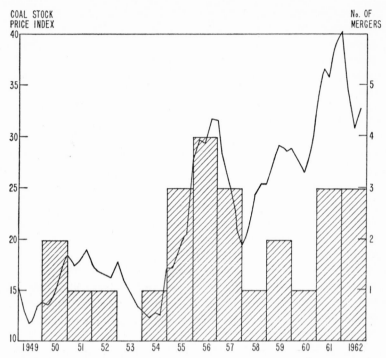

FIGURE 7. Relation of merger activity to coal stock price index, 1949-1962

each quarter in the period 1949–1962. Quarterly prices for Ayrshire Collieries, the fifth company included in the index, not being available, the average of its low and high prices for the first weeks of January, April, July, and October was used instead. These average prices for the five companies were then combined into a composite, unweighted average represented by the solid line in Figure 7. This index furnishes a better guide to midwestern coal company stock values than that provided by an industry index or the Dow–Jones Industrial index.

The chart indicates the existence of a fairly close correlation between merger activity in the midwest and the level of midwestern stock prices, particularly during the early years. In the mid-1950's when stock prices reached a cyclical peak, the number of mergers also attained its maximum. In the bull market from

1955 to the spring of 1957, ten of the twenty-six mergers consummated during the thirteen years for which we have merger data occurred. Merger activity reached its nadir when stock prices were lowest in 1953 and 1954. The one merger occurring in the latter year took place in December after stock prices had recovered from their early 1954 lows. Note that when stock prices recovered again in 1959 and 1961, merger activity increased as well, but with less vigor than previously. A point of diminishing returns appears to have been reached. Most of the likely candidates for merger had been involved in previous mergers, and few remained to be absorbed. For this reason the correlation between stock prices and merger activity probably will continue to be less pronounced in the future than it was between 1950 and 1958.

The most interesting question concerning midwestern coal mergers is the hardest one to answer. What were the motives for merger? Were they predominantly initiated by sellers anxious to leave a depressed industry? Did estate tax and investment considerations strongly influence sellers? Were buyers seeking to increase their market domination? Or did mergers merely offer buyers a relatively cheap and quick way to expand? Answers to these questions would require probing the minds of the participants, and this is obviously impossible. There are indications, however, that all of the above factors were at work in varying degrees in the twenty-six mergers covered in this study. Another thing is clear: the deft hand of the promoter is missing from these merger discussions.

Sellers took the initiative apparently in the Maumee–Peabody, Little Sister–Truax–Traer, Sherwood–Templeton–Pioneer and Minnehaha–Ayrshire mergers. Overtures may have been indirect; nevertheless these were instances in which sellers sought buyers. Estate and investment considerations were important factors in these cases. The Northern Illinois–Peabody, Consolidated–Bell and Zoller, Bradbury–Bell and Zoller, Chicago, Wilmington, and Franklin–Freeman, Binkley–Truax–Traer and Pyramid–Truax–Traer mergers were very likely initiated by sellers as well. In all but the Bradbury–Bell and Zoller mergers, the principals on the selling side were approaching retirement age and presumably took advantage of attractive offers to liquidate their investments.

In the remaining cases either buyers took the initiative or sellers were anxious to sell only if a sufficiently high price were established to make the sale attractive. Control of United Electric by Freeman obviously stemmed from the actions of the buyer since it was achieved by buying shares of stock on the open market. Consolidation Coal Company acquired Truax–Traer to gain a foothold in the midwest and to participate more fully in the growing electric utility market. Its previous experience had been with eastern operations. We are uncertain who initiated the purchase of Nashville Coal Company by West Kentucky Coal Company, but there is a strong suspicion that the buyer took the lead. At the time of its acquisition, Nashville was the largest nonunion producer of coal in western Kentucky. The close friendship between John L. Lewis, then President of the United Mine Workers, and Cyrus Eaton, Chairman of the Board of West Kentucky, and the aid of the U.M.W.A. in financing the purchase of Nashville have led industry observers to believe that Lewis was the prime mover in this merger. Shortly after the merger, incidentally, all the old Nashville mines were unionized. Whatever the motives for this merger were, it proved beneficial to the midwestern industry by eliminating a competitor which had expanded its production by vigorous price cutting. In fact, Nashville's disappearance marked the close of a period of price competition more intense than any experienced by the midwest in the post-World War II period.

Most of the Peabody acquisitions appear to have been aimed at controlling desirable reserves and improving the company's marketing position. The Sinclair–Peabody merger gave the Sinclair interests, who assumed control of Peabody, a valuable inheritance: a thirty-year contract to supply a minimum of half of Commonwealth Edison Company's growing coal requirements. Following the Sinclair–Peabody merger, every Peabody acquisition but one has involved the acquisition of attractive strip-coal reserves. The only exception was the purchase of Perry Coal Company which operated an underground mine. This was part of a joint purchase arrangement, the other company acquired being the Midwest–Radiant Coal Company which owned strip-coal reserves strategically located adjacent to some Peabody strip acreage. In the case of the Wasson, Forsythe–Carterville, and Hart & Hart acquisitions, Peabody never operated the mines. It acquired the companies solely to add to its strip-coal reserves. The same motive led to the merger with Duncan. These properties too were shut down immediately following their acquisition, but Peabody has used Duncan's strip reserves to expand its western Kentucky operations.

There is little evidence that coal mergers have achieved operating economies that could not have been attained in the absence of mergers. Technological economies of the type suggested previously by Penrose have followed some mergers. Production has often increased, following the introduction of larger cost-reducing equipment. But most of this equipment was within the financial reach of acquired companies if they had chosen to expand. Possibly minor managerial economies have flowed from some mergers from the "spreading" of overhead costs. As Stigler indicated, however, these savings usually are quantitatively unimportant. Thus, buyers have benefited from mergers by acquiring mining properties at less than the cost of building new facilities, by gratifying "empire-building" desires, by speeding up the expansion process, or by tightening market dominance through the restriction of competition. In some instances more than one of these factors may have been at work. There is no evidence of a conscious policy to achieve a fourth benefit—the limitation of competition; but such limitation has assuredly resulted from the increasing concentration of production. In an industry long plagued by destructively competitive practices, this is not necessarily a socially undesirable development. We reserve for the final chapter an analysis of this important problem.

Vertical Integration in the Midwestern Industry

The extent of vertical integration in an industry is an important aspect of its market structure that can influence industry behavior. This is certainly true, for example, in the petroleum industry where certain integrated producers control each step of the production and distribution process from crude-oil extraction to final distribution of the refined product at the retail level. The nature of coal production limits the ability of firms to integrate the production process. The only opportunity for vertical integration lies in tying distribution and production together. Midwestern operators have done little to take advantage of this possibility.

A few operators have attempted to control the transportation function, but this is not a widespread practice. Consolidation (through its subsidiary, Truax–Traer) and United Electric jointly own a barge line that moves their Illinois coal to river markets. West Kentucky Coal Company and Peabody also operate barge lines. Peabody operates two lines, one for general use along midwestern rivers; the other, jointly owned with Tampa Electric, hauls coal southward to a Tampa generating station which burns Peabody's Kentucky coal under a long-term contract. These coal companies operate barge lines not so much to economize on transportation costs as to insure themselves dependable barge service.

Midwestern operators rely almost exclusively on common carriers to move their rail coal to market; however, there are several exceptions. Peabody and Ayrshire jointly own the Yankeetown Dock Company which hauls coal a short distance from their two southern Indiana mines to loading facilities on the Ohio River for transshipment via barge. This short-line private railroad was a creature of necessity, there being no other means available to the companies to transport their coal to the river. Ayrshire constructed another short private rail line in western Indiana to connect Indiana and Michigan Electric Company's new Breed generating station to Ayrshire's Thunderbird mine which will supply the utility's entire coal requirements for fifteen years.

Midwestern producers have made only a few isolated attempts to control the final link in the distributive chain. For many years Peabody has owned the Crerar–Clinch Coal Company in Chicago, an operator of a river dock along the Illinois River and two retail yards within Chicago serving the local commercial and industrial trade. But it has not attempted to expand this operation in Chicago nor to introduce the idea elsewhere. It has also purchased the Michela Coal and Dock Company, an operator of coal docks at Marinette and at Washburne, Wisconsin, but this move was aimed at capturing a market denied them because of the traditional dominance of eastern-controlled dock companies operating along the west bank of Lake Michigan. West Kentucky Coal Company also has moved into the retail coal business to a minor extent, operating retail yards in Louisville, Paducah, and Nashville. And a few companies sell coal at retail to itinerant truckers who purchase coal at the mine for their own use or for resale to communities near the mine.

Several factors have limited integration forward to the retail level. The shortage of adequate financial resources in an industry burdened with low profits has probably been a contributing factor. Furthermore, vocal coal retailer associations which have wielded considerable power in the past have successfully fought against coal operators expanding into retailing. In addition, the structure of ownership in retailing has reduced the need for operator control of the retail function. Until recent years there have always been enough retailers operating in any market to insure each operator adequate coverage of the market. Whatever slight incentive may have existed to integrate forward toward the market has diminished in recent years in light of the declining relative importance of coal retailing.

We have observed that fuel costs account for 34 percent of total electric utility operating costs. In view of the great importance of a dependable low-cost fuel supply to the utilities, should they not integrate backstream by operating their own coal mines as the steel mills do?

Backward integration offers distinct advantages. First, it provides continuity of operation that lowers production costs. Steel companies, securing two thirds of their coking-coal requirements from "captive" mines, can operate their mines regularly, calling on commercial operators to take care of their marginal requirements. But for the steel mills there is an even more compelling reason for controlling their fuel supplies. Coal, in the form of coke, enters importantly into the ironmaking process; however, being more specialized than all-purpose steam coals, metallurgical coal deposits are correspondingly more scarce. Thus for the steel mills a second advantage of backward integration is to provide themselves adequate coking-coal reserves and to prevent artificially elevated market prices for this raw material. The steel companies accomplish this by securing their own reserves and by operating their own mines.

What is the situation with the electric utilities? Certainly the importance of fuel in the utilities' operations is as great as it is in ironmaking. Yet in 1955 only 11,000,000 of the 140,000,000 tons consumed by electric utilities in the United States came from "captive" operations. Backward integration in the midwest is even less prevalent. Louisville Gas and Electric Company, alone among all of the utilities in the six-state midwestern market area, operates a "captive" coal mine producing less than 400,000 tons of coal annually.

Four factors contribute to the reduced use of backward integration by the utilities. First, unlike the steel mills, utility plants can handle coals with diverse characteristics. Being less particular than steel companies about the quality of their coal removes the danger of reserve shortages and inflated supply prices in utility plants. Also, some utilities may have failed to integrate backstream through a lack of foresight. Prior to the post-World War II period when the coal industry was in the doldrums the average mine price often failed to cover average costs. Utilities would have been hard pressed to mine coal at costs below prevailing market prices. Today, faced with sharply increased coal requirements making vertical integration more feasible than before, most utilities would be equally hard pressed to acquire coal reserves sufficiently attractive to permit "captive" mine costs to compete with prices of coal sold in the commercial market by high-volume, low-cost producers.

Further, until recently few utilities consumed enough tonnage to warrant a mining operation of optimum size. Finally — and probably the overriding consideration in most cases — is the prevailing view in the electric utility industry that "the cobbler should stick to his last." Fear of possible unfavorable reaction by regulatory bodies if this industry moves into alien operations doubtless contributes to this view.

Degree of Buyer Concentration

This chapter so far has pointed up the degree of seller concentration existing in the midwest. Now we need to study the structure of coal buyers to assess the influence of their sizes and numbers on coal prices and, further, to determine whether market power rests primarily with the seller or with the buyer.

Data are available on two of the three major midwestern market segments: the electric utility industry and the retail trade. These markets are poles apart in terms of buyer concentration, there being large numbers of small retail dealers on the one hand, and utilities being oligopsonistic on the other. The other major market category, general industry, for which no statistics are available, falls somewhere between retailers and utilities in a scale of concentration.

Table 22 indicates the low level of buyer concentration in coal retailing. Even after their ranks had been severely depleted following the post-World War II decline in coal retailing, there were still 4,632 retail dealers in 1957 in the midwest's base market area.

TABLE 22. Concentration of retail coal dealers, selected states, 1957

State	Number of dealers	Tonnage (thousands)	Average tons per dealer
Illinois	1,787	8,624	4,825
Indiana	1,286	2,794	2,180
Iowa	603	1,252	2,076
Wisconsin	742	1,458	1,965
W. Kentucky (33 western counties)	40	160 (est.)	4,000
Missouri (34 eastern and southeastern counties)	174	700 (est.)	4,023
Total	4,632	14,988	3,235

Source: Comparative Fuel Costs, Bituminous Coal Institute, Washington, D. C., 1958.

Each dealer purchased an average of only 3,235 tons a year, barely more than a carload a week.[15] With few exceptions, dealers are too small, relative to the size of producers supplying them, to exert strong buying power over them. Also, most sellers' knowledge of the market is generally superior to retail buyers', and this adds to the retailers' disadvantage. Moreover, sellers often can strengthen their bargaining position vis-à-vis the retailer by directing advertising at the eventual consumer, tying him to the producers' coals. If the advertising is successful the dealer is virtually powerless to bargain effectively with the seller.

It is difficult to separate the influences pushing up the prices of retail coal above those prevailing in other market segments. That they are in fact higher than utility and industrial prices can not be denied. There are other factors than the level of concentration, however, generating their influence. Important among these are the seasonal nature of retail sales and the high costs involved in selling a fragmented market. These two factors probably contribute more than any other to the elevation of retail prices above the general coal price level. Nevertheless, the superior bargaining position of the seller, though it cannot be measured, doubtless contributes to this phenomenon as well.

The concentration of buying power in the electric utility industry is diametrically opposed to that existing in the retail trade. Here, instead of thousands of small firms buying a few tons of coal a week, twenty-four companies purchase vastly greater amounts, in what can accurately be described as oligopsonistic markets. Table 23 lists the principal utilities in the midwest's market area, together with their 1959 coal consumption. In that year they consumed 43,250,000 tons of coal of which 41,364,000 tons came from mines in the Eastern Interior region. They accounted for over 75 percent of the midwest's shipments to utilities in 1959.

Such a high degree of buyer concentration obviously increases the utilities' market power over what would prevail if buyers were more numerous. The extent of this market power, however, is not uniform. Competition among shippers at various utility plants varies according to the number of producers capable of supplying

---

[15] A carload represents a minimum shipment of coal by rail.

TABLE 23. Consumption by largest midwestern electric utilities, 1959

State	Company	Coal consumption (thous. tons)
Illinois	Commonwealth Edison[a]	8,699
	Electric Energy Inc.	3,296
	Union Electric[b]	2,918
	Illinois Power	1,975
	Central Illinois Public Service	1,341
	Central Illinois Light	865
	Central Illinois Electric and Gas	257
		19,351
Indiana	Indiana and Kentucky Electric Corp.	5,828
	Public Service Company of Indiana	2,659
	Indianapolis Power and Light	1,414
	Northern Indiana Public Service	850
	Southern Indiana Gas and Electric	524
		11,275
Kentucky	T.V.A.	4,835
	Louisville Gas and Electric	1,152
	Kentucky Utilities (Green River plant only)	542
		6,529
Wisconsin	Wisconsin Electric Power	2,688
	Wisconsin Public Service	743
	Wisconsin Power and Light	706
	Dairyland Power Cooperative	420
		4,557
Iowa	Iowa Electric Light and Power	522
	Interstate Power[c]	426
	Iowa Power and Light	287
	Iowa–Illinois Gas and Electric[d]	200
	Iowa Public Service	103
		1,538
Grand total		43,250
Total (excluding consumption of coal from outside states)		41,364

[a] Includes 1,031,000 tons consumed by Commonwealth Edison in Indiana.
[b] Includes 1,354,000 tons consumed by Union Electric in Missouri.
[c] Includes 16,000 tons consumed by Interstate Power in Minnesota and 46,000 tons in Illinois.
[d] Includes 29,000 tons consumed by Iowa–Illinois Gas and Electric in Illinois.
Source: Steam–Electric Plant Factors, 1959, National Coal Association, Washington, D. C., July, 1960.

the plants. There is no price competition more severe than that prevailing in the Tennessee Valley Authority's markets. Several factors contribute to this condition, not the least of which is the fact that T.V.A.'s plants are strategically located near coal fields and on navigable streams. Its southern and eastern plants are situated in an area where coal production is unconcentrated, which affords T.V.A.'s buyers a unique opportunity to play off one small seller against another. In other regions, however, some utilities, dependent on a few producers for their coal, possess far less market power. For example, Central Illinois Light and Power Company's plants located near Peoria must rely for their coal supply on the few mines operating in the nearby Fulton–Peoria and Springfield districts. The same holds true for Illinois Power Company's Havana, Illinois, plant. Both companies could draw on coal supplies from more distant districts but only by paying substantially higher delivered prices. Coal shippers in the mining regions adjacent to these and other similarly situated plants, therefore, possess an important locational advantage that partially nullifies the utilities' oligopsonistic market power.

There is an additional, highly significant, factor counterbalancing the utilities' apparently dominant market position. Just as coal companies must increasingly rely on sales to utilities for survival, so too are the utilities dependent on coal producers for reliable long-term coal supplies. As their coal requirements expand, utilities are becoming increasingly aware of this mutual dependence. To insure themselves satisfactory fuel supplies all large utilities contract for substantial portions of their total coal requirements on a long-term basis. Coal operators demand and receive higher prices on sales of this type to justify making long-run commitments of their resources. As seller concentration increases, the balance of power should continue to move toward the seller as the oligopsonistic power of buyers is matched by the oligopolistic force of sellers.

CHAPTER V

Cost Aspects of Structure

This chapter continues a study of market structure traits having a bearing on conduct and performance in the midwestern coal industry. We turn our attention to production conditions within the industry to learn the influence of resource and operating conditions on cost levels. The influence of coal depth, for example, determining whether an operation uses the strip- or underground-mining method, has an important bearing on costs and on the character of competition.

Studying the relation of fixed to variable costs increases our understanding of the effects on earnings of changes in demand. Moving beyond short-run costs, we also need to study returns to scale for the bearing they have on entry barriers (Chapter VI). Finally, in light of the dramatic recent advances in productivity in coal, we analyze technological change as a structural influence.

Methods of Extraction and Coal Preparation

The methods used in mining coal influence enough important economic factors to warrant summarizing the techniques involved. They directly affect production cost levels and govern the efficiency with which resources are utilized.

Coal mining may be carried out either below or above ground. In underground mining, coal is brought to the surface in one of three ways: through use of a shaft, slope, or drift (horizontal) opening. Topographic conditions and equipment used determine which of the three systems is adopted. The openings are used to convey men, supplies, and equipment to and from the working areas and to transport raw coal from underground workings to the surface.

In a conventional, mechanized, underground mine, mining is divided into several distinct operations. The coal is first undercut to facilitate its breaking away from the solid face. The seam is then drilled, these drill holes are filled with explosives, and a section of the coal is blasted away from the seam. Mechanical loading machines gather in the shot-down coal with clawlike arms, pulling it onto a built-in conveyor. The coal discharges from the conveyor into a waiting shuttle car or a longer extensible conveyor which transports it to the main haulageway. From there, the coal moves via either another conveyor or an underground, small-gauge railroad directly to the surface, or in the case of a shaft mine, to a skip hoist which subsequently lifts the coal above ground.

Though this system is far removed from former techniques in which coal was hand loaded and mules and horses provided the energy for underground transportation, still it is a laborious and costly method of extraction. In recent years, continuous mining machines have been perfected which combine into one operation the separate jobs of undercutting, drilling, blasting, and loading.

Whether a conventional or continuous mining system is used, most underground mining is done by the room-and-pillar method. The area underground is crisscrossed with haulageways and entries which divide the mine into separate segments. Each segment again is divided by entryways into workable areas, referred to as rooms. Barriers and pillars of coal are left as roof supports between rooms and around the main shaft and haulageways to prevent subsidence. The loss of coal in these supports reduces the coal recovery in most underground mines to 45 to 60 percent of the original reserve.

Each room is a separate production center, outfitted with a complete set of machines and connected with haulageways by its own minor transportation system. The older a mine becomes, the farther from the main shaft these production centers move. Underground transportation costs increase concomitantly. Diminishing returns in underground mining arise principally from the wavelike expansion of production farther and farther away from the shaft or slope opening.

In addition to underground mining, an increasing proportion of total production today comes from surface, or strip, mining. The strip-mining production cycle differs sharply from the methods used in underground production. The consolidated and unconsolidated material (called overburden) lying above a coal seam is first removed by shovels or draglines whose bucket capacity ranges in size from a cubic yard or two to 200 cubic yards on the behemoths currently going into production. The coal having been laid bare by the excavating equipment, smaller shovels then load it into large trucks to be hauled via mine roads to the preparation plant for processing.

Strip mining offers decided advantages over underground in several respects. First, mining is conducted in the open, eliminating the need for ventilation and increasing the safety factor. Moreover, the use of large, mechanized equipment reduces labor requirements and cuts costs. Finally, removing the requirement for support pillars and barriers conserves coal by increasing coal recovery.

Distinctions between underground- and strip-mining operations disappear when the raw coal enters a preparation plant. In small, marginal operations the preparation plants are replaced by crude jerry-built loading ramps used to load coal into railroad cars. In an increasingly large number of modern mines, however, coal receives additional treatment before being shipped to market. Basically a simple operation, the preparation plant first washes the coal to remove bedded impurities and foreign matter introduced in the mining process. The coal is then sized and in some plants dried, as a freeze-proofing measure, before it is loaded into open-top cars. If the demand for smaller sizes exceeds the amount created naturally in the mining process, larger sizes are crushed to the appropriately smaller dimensions.

Effect of Resource Conditions

Many factors influence the character of operations for each coal mine. Depth of cover over the seam to be worked is extremely important. If the coal is shallow — at depths usually less than 100 feet — stripping is called for. Deeper seams must be mined by underground methods, and the deeper the seam, the more problems encountered in the operation. Costly shafts or slopes must be dug deeper, and roof support problems multiply with the increasing pressures exerted by the overlying strata.

Thickness of the coal seam being worked is a crucial factor. In strip mining, stripping costs vary inversely with seam thicknesses. A stripping unit capable of uncovering 50,000 tons of coal per month with a three-foot seam will uncover 100,000 tons with a six-foot seam at no additional expense. Advantages of thick seams in underground mining are equally important. Working places are less cramped; larger mine cars can be used; haulage distances are reduced per ton of coal extracted; and fewer working centers are required to produce a given daily tonnage, thereby reducing supervisory costs and improving coordination.

The character of the roof above the seam and the material directly below the coal are particularly important in underground operations. Though core drillings give the operator advance indications of roof conditions, he is never certain how well the roof will support the pressure from above until the mine investment is made and operations commence. The hazards of faulty roof conditions and seam irregularities (underground faults and pinched-out seams) introduce more risk into underground operations than is generally found in strip mining.

Above ground, the topography of the terrain over the coal is influential in several respects. In the flat country typical of the midwest, roads, preparation plants, and auxiliary mine buildings are more readily constructed than they are in the West Virginia and eastern Kentucky hill country. There, mines wedge themselves into cramped valleys carved out of the steep hillsides. The topography in the hilly eastern mining regions, however, provides mines in that region with one distinct advantage over midwestern operations: the ability to reach underground seams with drift openings, less costly to construct than the shafts and slopes which the flat midwestern landscape dictates.

In most important respects, however, mining conditions in the midwest are superior to those in the east and elsewhere. Seams on the whole are thicker and more level, lending themselves to large-scale mechanized operations. Moreover, a great proportion of the coal can be won through the use of large stripping equipment operating on extensive level stripping areas. Whatever open-cut mining is done in the east is conducted by contour strippers cutting into the sides of mountains. The ability to mine a large percentage of its coal by the efficient stripping process gives the midwest a decided advantage in productivity over its rival coal fields.

Short-Run Costs

Anyone measuring costs at various output rates encounters problems that are not easily resolved. Not least of the problems is the difficulty of securing data. Most detailed cost data are securely locked up in company vaults, inaccessible to the interested investigator. Published data are confined principally to trade association reports, but they are deficient in several respects. The investigator using time-series figures is unable to differentiate between changes in costs resulting from different rates of output, varying mining conditions, and fluctuating factor prices.[1] The only satisfactory way to correct these deficiencies is to construct hypothetical cost curves based on actual costs, projecting costs at different output rates assuming a given division of fixed and variable costs. This we do in Figures 8 and 9 for strip and underground mines, respectively.

Dividing costs into their fixed and variable components is not easy. J. M. Clark in his definitive work on overhead costs recognizes this:

> Starting with a search for certain accounting items which do not vary at all with variations in business and other items which vary in proportion to business, one soon finds that there are no items that remain permanently unchanged, few that remain unchanged for relatively short periods, and none that always vary exactly in proportion to business. As a result, every item of cost is bound to have a mixed character.[2]

Some studies label fixed costs as those that continue at a zero output rate. But this interpretation raises problems since fixed costs may vary depending on the length of time output is halted

[1] The only attempt we have seen to relate costs to output rates for the coal industry using time-series data reveals an anomaly — decreasing operating costs per ton at declining operating rates over a small range far short of capacity output.

[2] J. Maurice Clark, *Studies in the Economics of Overhead Costs* (Chicago: Univ. of Chicago Press, 1923), p. 51.

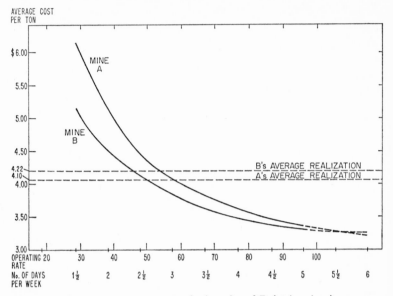

FIGURE 8. Short-run average costs of mines A and B (strip mines)

and, indeed, may change because of differing expectations about the length of the shutdown. Clark, arguing that these costs might better be called "minimum" or "shutdown" costs, prefers to study cost-output relationships merely by viewing the effects on total costs of varying output.[3] This approach, however, fails to divide costs into fixed and variable components, and is, therefore, useless for constructing a hypothetical cost schedule. In the present study an arbitrary division classifies as fixed costs those that remain unchanged over various *positive* levels of output.

In our calculations fixed costs include: depreciation, interest, taxes, insurance, administrative and supervisory salaries, miscellaneous administrative expenses, and selling costs. The underground-mine cost data include a fixed dollar amount to cover sales costs. The strip figures arbitrarily include 16 cents per ton times the tons produced on a 4-day week. The 16-cent figure equals the average sales cost per ton for Indiana strip mines in 1956 (the latest authentic data we have on these costs). Four days

[3] *Ibid.*, p. 54.

COST ASPECTS OF STRUCTURE

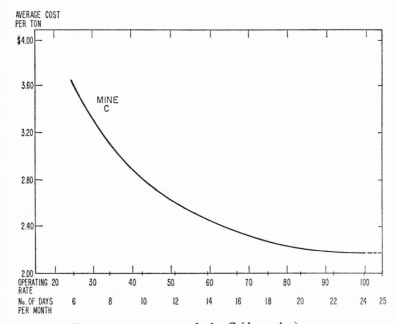

FIGURE 9. Short-run average costs of mine C (deep mine)

per week approximates the average work week for midwestern strip mines. The figures are derived arbitrarily because using the actual figures to which we have access would be misleading. One of the strip mines operates a retail coal dock which raises its sales cost far above the average for the industry selling coal on a wholesale basis; the other mine contracts to have its coal sold by a sales agent, thus converting this expense into a variable cost. In the usual case, however, the larger firms manage their own sales force with most expenses remaining fixed (except for some sales commissions) over wide output ranges.

Figure 8 plots average total costs for two medium- to large-sized Illinois strip mines based on 1958 costs.[4] In Figure 9 the

[4] The company operating these mines furnished detailed cost sheets with the figures sufficiently broken down to distinguish between fixed and variable costs — except for a few minor items such as power costs for which the company provided no breakdown. The two mines are in the 450,000- to 750,000-annual tonnage range.

underground mine costs are based on hypothetical cost data submitted by Peabody Coal Company during hearings studying the extension of the Guffey Act. They project costs at various output rates for a large Illinois underground mine operating in 1941.[5] Costs for both modes of mining are necessary in view of the importance of strip mining in the midwest, accounting for approximately 55 percent of the Eastern Interior region's output.

The cost curves display similar characteristics. Mine A's cost curve lies above B's to the left of the five-and-one-half days per week output level, reflecting its higher fixed costs. B is a comparatively old mine, burdened with no interest costs and low depreciation charges, unlike A with its fairly high interest and depreciation costs. Mine C's cost curve rises less steeply than either of the strip-mine curves, its flatter slope being a function of the mine's lower overhead costs. At 50 percent of capacity, the mine's costs are 20.4 percent above those incurred at full capacity operations; at this output rate, B's costs increase 24.2 percent and A's 31.5 percent. But all three curves exhibit generally similar slopes, flattening out as they approach capacity outputs. Dotted lines project costs beyond theoretical maxima, 260 operating days for the strip mines and 280 days for the deep. This extension may represent doubtful extrapolation for strip operations whose outputs are rigidly determined in the short run by stripping equipment which must work continuously to uncover enough coal for 260 operating days. More than likely, the strip cost curves rise very steeply, perhaps vertically, when capacity is reached.

It is difficult to determine whether the average cost curves in Figures 8 and 9 fairly depict average industry costs. Since they cover costs for only three mines operating in two past time periods under varying mining conditions, the absolute cost levels cannot be considered representative of existing average costs. But the curves' *shapes* probably represent fairly the cost functions of modern, large-scale, mechanized mines which supply an increas-

---

[5] The difference in absolute costs for the 2 mining methods results largely from the fact that costs covered 2 time periods. If the underground mining cost curve reflected 1958 costs it would doubtless lie above both strip mine cost curves. The ordinates and abscissae on the 2 graphs are drawn approximately to the same scale to permit visual comparison of the curves' slopes.

ingly large share of today's coal requirements.[6] Smaller, less mechanized mines with lower capital/output ratios doubtless have flatter curves, reflecting their lower depreciation charges. Wartime Office of Price Administration's cost records provide a benchmark for comparing our data with industry averages. In 1944, when the midwest operated at approximately 94 percent of capacity, mechanized deep mines in this region had fixed costs averaging 16 percent of average total costs; midwestern strip mine fixed costs stood at 25 percent of total costs.[7] The fixed cost component of mine C's costs account for approximately 21 percent of total costs, lower than the 25 to 33 percent figure for mines B and A. The slightly higher level of overhead costs for the three mines probably is a function of their greater than average mechanization. Fixed costs may be one or two percentage points on the high side (both for the O.P.A. data and for our strip-mine hypothetical cost curves) from the inclusion of several cents per ton sales commission expense that legitimately belongs under variable costs. This error is countered somewhat by failing to divide power costs into fixed and variable components. Since both the O.P.A. figures and ours are similarly treated, the errors (slight as they are) affect only the absolute level of fixed costs and not the relationship of fixed to total costs, our data vs. the O.P.A.'s.

Average variable costs — hence marginal costs — for strip mines are virtually constant over a wide range of output. Output for both strip and underground mines fluctuates in the short run through manipulation of the number of days worked per week. But productivity per day varies imperceptibly when the work

[6] The U.S. Coal Commission [*Report on the Effect of Irregular Operations on the Unit Cost of Production* (Washington: U.S. Government Printing Office, 1925), Pt. 8, p. 1975] reported that, taking 25 days a month to represent full-time operations, average costs per ton rose 8 to 9 percent when the mine operated 16 days, 21 to 25 percent for 12 days and 48 percent for 8 days. On a comparable basis, the Mine C figures increase 8.6 percent, 20.1 percent, and 43.8 percent, agreeing pretty closely with the Coal Commission's national averages.

[7] Based on cost breakdowns in: Office of Price Administration, *Preliminary Survey of Operating Data for Commercial Bituminous Coal Mines for the Years 1943, 1944, 1945,* O.P.A. Economic Data Series no. 2 (Washington: U. S. Government Printing Office, n.d.), pp. 17, 23. Data include all commercial mines producing more than 50 tons per day.

week shortens. Average variable costs rise as the work week lengthens beyond five days, reflecting the payment of time-and-one-half wages for overtime; but average total costs continue to fall from the downward pressure of lower average fixed costs.

Underground-mine average variable costs, too, tend to remain constant in the high-output ranges; however, they apparently rise somewhat with sharp output reductions. A study of Indiana underground-mine productivity indicates that when the work week declines below the four-and-one-half-day per week level, output per man-day falls noticeably, and is accentuated when the mine operates only two days per week.[8] Mines operating eighteen to twenty-two days per month report production of 8.06 to 9.50 tons per man-day. Those working 11.75 to 16.75 days per month witness a decline in man-day output to 7.18 to 7.61 tons; at nine days per month, the figure falls to 6.70 tons. The result is not surprising in view of the tendency for underground mining conditions to deteriorate rapidly when operations cease temporarily. Roof falls on idle days present a cleanup problem that delays production when the mine resumes operations. Costs consequently increase.

What significance can we attach to the short-run average cost curves? First, assuming the continuation of existing mines, technology, and plant scale, cost reductions should decelerate if the midwest continues to reduce excess capacity. At 80 percent of capacity (the approximate current midwestern operating rate), the two strip-mines' costs lie only 5.8 and 8.5 percent above projected costs at the capacity output level, underground-mine costs, only 3.5 percent. Relaxing the assumptions affects this cost-output relationship. Even if factor prices rise to absorb cost reductions from improved technology and increasing returns to scale, the changing mix of existing mines should increase efficiency as excess capacity continues to decline. In the long run, only the fittest should survive, the least efficient firms being the first to be driven from the industry. Much of the spectacular postwar rise in efficiency in the coal industry probably derives from this Darwinian principle. But even here the rate of efficiency improvement should decelerate as output approaches the capacity level.

[8] Harline, "Economics of the Indiana Coal Mining Industry," p. 251.

If the cost reduction rate decelerates with increased output it follows that increases accelerate when output falls. For each of the full-day increments between five-and-one-half and one-and-one-half operating days per week, for example, Mine B's costs per ton rise $.219, $.382, $.682, and $1.604, respectively. As output falls the incentive for each operator to cut prices should increase. Each operator seeing the demand curve facing him as being elastic feels that this course of action commends itself to him. The twin conditions of rapidly rising costs resulting from high overhead charges and the use of price reductions by each operator to stimulate demand for *his* product allegedly have lead to the destructively competitive conditions that have long characterized the industry. Each of these reactions needs studying.

The literature on the coal industry appears universally to accept the notion that fixed costs in coal mining are high.[9] This belief may be partly an illusion growing out of conditions existing during the 1920's and 1930's when the observation was often made. Throughout much of that period, the nation's mines operated far below capacity, and when this condition occurs fixed costs, as a percentage of total costs, are bound to rise. Observe costs during prosperous times and the fixed cost percentage falls. An assertion, therefore, that fixed costs are high is meaningless in the absence of assumptions concerning the operating rate. Furthermore, the claim that overhead charges in coal are inordinately high is clouded by the existence of extraordinarily high labor costs in mining. Since most labor costs enter into the variable cost component, it is hard to reconcile these apparently conflicting notions.

The problem really centers on an unambiguous definition of the word "high." Fixed costs that one observer may view as high may appear to be moderate to another. J. W. Markham in his study of the rayon industry considered fixed costs in this industry to be high, accounting for 22 and 33 percent of total costs.[10] Using this

---

[9] Typical is Parker's assertion: "Coal operators are very sensitive to changes in the volume of production since there is an unusually large element of overhead. Capital charges go on unabated, and as days of operation diminish, the cost per ton increases rapidly." Glen Lawhon Parker, *The Coal Industry* (Washington, D.C.: American Council on Public Affairs, 1940), p. 6.

[10] Jesse William Markham, *Competition in the Rayon Industry* (Cambridge: Harvard Univ. Press, 1952), p. 150.

criterion, fixed costs for the three mines in Figures 8 and 9 are also high, attaining approximately the same level.

Two additional aspects of the relationship between excess capacity and pricing practices remain. The first derives from the previously noted tendency of the rate of cost increases to accelerate in the short run as excess capacity increases. If the operating rate stays reasonably high — say above the 70 percent level — costs do not rise appreciably above the low level achieved at capacity operations.

The final factor that needs emphasizing is the absurdity of translating the adverse cost effects of increased excess capacity automatically into immediate and direct effects on the price level. Many coal prices are protected in the very short run through the use of contracts binding seller and buyer. Moreover, prices may lie sufficiently far above average costs to provide a comfortable profit margin even at fairly low operating rates. The price-cost relationships in Figure 8 reveal that the mines broke even at an operating rate in the vicinity of 50 percent. It could be argued, correctly, that this condition would not hold if output of all firms were reduced since such a general reduction in demand would depress prices, thereby raising the break-even point.

In addition, the assumption that high fixed costs spur price reductions when output falls fails to account for different market structures that can modify the results. When many sellers scramble for a small share of a common market prices should respond readily to demand changes, in conformance with the purely competitive model. This conclusion is modified if all or most short-run cost curves are horizontal or approach horizontality and have identical or closely similar positions, and if the industry demand curve shifts within the relatively horizontal portion of the short-run industry supply curve. With these assumptions price would vary little if at all. When sellers are few, the response of each may be different. The "rules of the game" may call for stable prices even in the face of fairly sharply reduced demand. The unresponsiveness of price to curtailed demand in the cement and steel industries supports this view. Increased seller concentration in the midwestern coal industry might well lead to the adoption of a similar pattern.

Returns to Scale

In the long run the firm may increase output by expanding existing plants, by replacing existing plants with ones capable of greater output, or by adding plants to those already in operation. Supposedly, economies flow from these increases in the scale of operation, at least up to a point. Scale economies may take two forms: they may stem from increases in the scale of plant or from increases in the size of the firm.

As plants and production processes grow larger, factors usually join in more efficient combinations. The principle of the division of labor increases labor efficiency by permitting job specialization. Machinery efficiency improves from what E. A. G. Robinson calls the "integration of processes," that is, the use of one machine to perform several jobs.[11] This phenomenon is present in coal mining in the use of continuous mining machines which combine the traditional functions of drilling, shooting, and loading into one operation. The indivisibility of some machines permits economies from the use of large machines as plant scale increases. Again, in coal, a case in point is the use of increasingly larger stripping units in large-scale strip mines. Finally, specialization carries over into management as increasing the plant scale permits a narrower division of managerial functions.

Internal economies for the firm come from a number of directions. Managerial economies allegedly increase from the further specialization of management functions. Factor costs may decline as a result of economies from large-scale buying. Financial economies presumably flow from the ability of larger firms to command capital funds at lower costs. And the firm may reduce marketing costs by using mass distribution techniques.

A study of long-run costs is necessary for several reasons. We need to know the extent to which firms can achieve scale economies to understand whether enlarging productive units in the industry serves industrial efficiency. And on the other side of the coin — we must determine if the optimum plant and firm sizes are so large that the penalty for maximum efficiency is a severe contraction in the number of sellers in the industry. If long-run cost

[11] Robinson, *The Structure of Competitive Industry*, p. 19.

curves are negatively sloped over wide-output ranges, there may be a built-in tendency toward oligopoly in the industry.

Scale economies for the firm are harder to measure than are those for individual plants. Also, there is some question concerning the extent to which they in fact exist. Managerial economies do not readily lend themselves to measurement. The room for cost advantages in marketing, financing, and labor and material purchasing is severely limited. Larger firms possess sales advantages not enjoyed by smaller firms, but these are mostly qualitative. Firms large enough to have their own sales departments gain a pecuniary advantage over smaller firms forced to rely on brokers and sales agents, but the advantage is slight.

It is alleged that larger firms can reduce capital costs by borrowing money at cheaper rates, but even here the advantage is probably negligible. Compare the interest charges for a company borrowing money at 5 percent, 16 2/3 percent less than its smaller competitor forced to pay 6 percent. The larger firm, erecting a one-million-ton per year plant costing $8 million (half of which we assume is borrowed) saves an average of approximately 2 cents per ton over the period of the loan liquidation; and taxes reduce this small saving by one half. The remaining item — labor and supply costs — offers little room for economies for the large firm. An industry-wide labor agreement determines wage rates, and individual unionized operators are powerless to change them. Larger firms do possess some leverage that reduces slightly the cost of some supply items, but in the aggregate these savings, too, are minor.

When we compare the productive efficiency of *plants*—as opposed to *firms*—we are on firmer ground since the available records permit quantification. The accepted industry efficiency criterion is a measurement of tons produced per man-day. It is subject to some limitations. The Bureau of Mines and most of the states' Divisions of Mines compute the figure for each mine by dividing annual tonnage by the average number of men employed times the number of days the tipple operates. This measurement disregards the fact that in strip mines some men work on idle days removing overburden and, in all mines, doing repair work. Moreover, it is deficient because it measures input in terms of days, instead of hours, worked. Though the tons per man-day concept probably distorts the absolute levels of productivity, it is, nonetheless, sufficient for the comparative purposes to which we put it.

Comparing productivity for both underground and strip mines of various sizes should provide a rough guide to the availability of plant scale economies. Table 24 presents measures of average

TABLE 24. Average output per man-day, by mine size, Illinois mines, 1959

Mine size (annual production, tons)	Underground mines (tons per man-day)	Strip mines (tons per man-day)
+500,000	19.2	37.7
50,000–500,000	19.0	31.2
10,000–50,000	7.7	20.4

Source: Annual Report, State of Illinois Department of Mines and Minerals, 1959.

productivity for large, medium, and small Illinois strip and underground producers for 1959. Economies of scale are more evident in strip than they are in underground mining. Productivity in underground mining jumps sharply as output expands from the 10,000- to 50,000-ton range to a production of 50,000 to 500,000 tons. But as output rises above 500,000 tons, plant scale economies apparently cease. The strip mine figures show a different pattern, with productivity continuously increasing as mine size expands.

We might legitimately ask: to what extent is the relationship between productivity and mine size attributable to varying mining conditions? In the case of strip mines especially, the improved performance of larger mines may stem not from the use of more productive techniques but from the use of more desirable coal reserves. To check this aspect of the problem, we computed average stripping ratios for all strip mines analyzed, and for the underground mines, average seam thicknesses and depths. Stripping ratios for the three groups, from the largest to the smallest, were 11.8:1, 9.3:1, and 5.4:1, strengthening the view that increased productivity for this mode of mining derives from the use of more capital-intensive techniques.[12] In underground mines seam thicknesses declined from the largest group to the smallest, from 7.6 feet to 6.3 to 4.9, but seam depths declined as mines grew smaller, averaging 310 feet for the largest, 230 for medium-sized mines and 219 for the smallest. The differences in depth are probably not enough to make an appreciable difference in costs, nor would costs be influenced much by the spread between seam thicknesses for the two largest groups. Since seam thickness falls below 5 feet for mines in the 10,000- to 50,000-ton class, they probably operate at a slight disadvantage compared with their larger competitors. Some of this group's lackluster productivity performance probably derives from its deficient reserves, but other factors account for most of the differences.

The slightly different scale-productivity pattern for underground and strip mines results from the nature of the two mining processes. There is more divisibility in underground mining operations than in strip. Medium-sized deep mines can duplicate the facilities of larger operations, the differences in total output being a function principally of the number of underground "places" worked. Since each face operation works semi-independently of others, the addition of face crews increases total tonnage without measurably increasing productivity. This phenomenon accounts for the inability of large mines to improve productivity as the scale of plant enlarges. The trend toward high-speed continuous miners may push up the minimum size of the optimum scale plant, but the minimum will probably continue to lie below the 500,000-ton mark. Many of the mines in the 10,000- to 50,000-ton class are less mechanized than the larger mines, and this fact plus inefficient management accounts for their poorer than average productivity performance.

The opportunities for scale economies in strip mining are limited only by an operator's financial resources and coal reserve position.

---

[12] This statement requires a word of caution. It would be erroneous to conclude that there is a direct relation between stripping ratios and productivity, that is, that low ratios lead to low productivity. In fact, the reverse is true. Given 2 mines operating *with the same equipment, in the same coal field but at different ratios,* the mine with the lower ratio should have the higher productivity.

Mines in the smallest size classification use one-and-one-half- to five-cubic-yard capacity stripping machines, small dump trucks, and drills; larger mines employ six- to eighteen-cubic-yard machines, larger trucks and drills; the biggest mines use twenty-five- to two-hundred-cubic-yard capacity shovels and draglines, ten- to twelve-inch, high-speed overburden drills and coal haulers with capacities of twenty-five to one hundred tons. As equipment size grows, productivity continues to increase. It is difficult to say what percentage of total industry output is represented by strip mines operating at minimum costs since rapid developments in equipment technology continually lead to larger mine sizes. It seems clear, though, that a ceiling is in sight. In the absence of unforeseen technological innovations, diminishing returns resulting from the expansion of production away from the preparation plant effectively limit the optimum plant size. Peabody Coal Company's four-million-ton-capacity mine in western Kentucky, the country's largest strip coal mine, accounts for less than one percent of the nation's coal output, less than 5 percent of midwestern production. Except in unusual circumstances, the feasible limit certainly lies not far beyond the size of this mine.

Three exceptions might contradict this assertion. The first would be the creation of larger mines that are "mobile" in the sense that all equipment — tipple, excavating equipment, trucks, etc. — are moved periodically to reduce haulage costs as mining moves away from the tipple. Second, strip mines might exceed this size appreciably if coal seam thicknesses are substantially greater than average. In the absence of these conditions mining proceeds so rapidly that haulage costs quickly become exorbitant. For example, a mine producing 4 million tons annually, working in a 5-foot seam removes over 500 acres of coal yearly. In 10 years nearly 8 square miles of coal will have been removed, moving operations in some instances up to 4 to 5 miles distant from the preparation plant. But the third possibility might mitigate this problem. Technology in coal haulage might keep pace with improvements in stripping equipment, permitting pit operations to move far from the tipple without unduly affecting costs.

Three conclusions stand out from the foregoing analysis. The first is that as the trend continues toward more large mines and fewer in the low tonnage class than before, productivity in both strip and underground mining should improve. But the ceiling on productivity is apparently higher in strip mining than in deep mining. This situation should continue until such time as there is a major technological breakthrough in underground mining. Second, peak efficiency in underground mining being reached at fairly low tonnage rates, barriers to entry *from the productivity standpoint* are low. The force of this conclusion is weakened, however, by the fact that at their most efficient, underground mines are half as productive as the most efficient strip mines.[13] Later we learn that there are other impediments to easy entry operating for both strip and deep mines in the form of coal reserve barriers. Finally, since optimum scale plants are comparatively small, and account individually for small shares of industry output, the trend toward concentration of ownership cannot be justified *on efficiency grounds*. Scale economies evidently are as attainable under an atomistic market structure as they are under an oligopoly.

Structural Influence of Technological Change

The pressures for technological change in the coal industry have been uneven both temporally and with respect to the different producing regions. Christenson distinguishes among four mining regions building up pressures for mechanization that set them apart from the rest of the industry. Illinois, Indiana, and western Kentucky make up one of these regions. Christenson views the pressure for mechanization in the midwestern region as resulting from that area's relatively low-rank coal struggling for survival against higher-quality eastern coal.[14] This thesis is plausible and superficially correct, but a deeper analysis reveals a more important underlying cause — the basis for which rests in the industry's labor relations' history. Indeed, the two are so

---

[13] This is not to say that their costs will be twice as high as strip since there are numerous cost items which are similar or identical (viz., welfare fund payments, sales costs, administrative overhead, etc.). Lower supply costs counteract part of the underground mine's labor cost disadvantage.

[14] Christenson, *Economic Redevelopment in Bituminous Coal*, p. 253.

closely intertwined that we cannot discuss one without an understanding of the other.

Illinois and Indiana, along with Ohio and western Pennsylvania, formed the nucleus of the United Mine Workers of America's bargaining strength, beginning with the Central Competitive Field Interstate Agreement of 1898. Except for minor defections this bargaining group remained intact until 1927, setting the pace for wage bargains reached in other unionized districts. Still, throughout the early part of this period many districts in West Virginia, eastern Kentucky, and Tennessee operated on a nonunion basis at slightly lower wage levels than those prevailing in the north. But a growing demand for coal minimized the competitive pressure of nonunion wage differentials. More important than wage competition as determinants of interdistrict competitive strength were location, coal quality, and productivity.

Following World War I, conditions changed swiftly. Coal demand languished as the country moved from war to peace. Pressed by competition from oil and gas, coal suffered further as it moved into a period of secularly stagnant demand. Southern nonunion operators in West Virginia, eastern Kentucky, and Tennessee met the challenge by cutting wages, widening the wage differentials between their operations and those in the northern union mines rigidly tied to the U.M.W.A.'s contract. In 1920 the basic rate for inside workers in northern mines was $7.50 a day, about 4 percent above the southern nonunion rate; within the next few years rates in the south had fallen to $3.50–$5.00 a day, while the north held the line at $7.50.[15] Adamantly, John L. Lewis, the Mineworkers' president, announced that the union would take "no backward step."

Reaction to the new demand and labor conditions was immediate and severe. The proportion of the United States' coal output produced in union mines declined from 72 percent in 1919 to 40 percent in 1925. Nonunion output accounted for 93 percent, 98 percent, and 90 percent of the production from West Virginia, eastern Kentucky, and Tennessee, respectively. Illinois and Indiana operators remained steadfastly loyal to the union, but the penalty they paid was high. Armed with lower direct labor costs,

[15] Harline, p. 85.

southern mines invaded northern markets, disturbing the competitive equilibrium and capturing large tonnages. In 1920 the average realization for Indiana coal was $1.18 a ton below West Virginia's average price; benefited by reduced costs, West Virginia's mine prices fell 9 to 34 cents per ton below Indiana's during the period 1924 to 1932.

The midwest's tonnage losses were widespread. Despite an increase of 75 percent in Michigan's coal consumption between 1917 and 1929, shipments into the state from Illinois and Indiana declined from 1,380,000 tons to 84,000 tons.[16] The states of Indiana, Illinois, Iowa, Wisconsin, and Minnesota — the heart of the midwestern industry's market — acquired only 47 percent of its coal in 1929 from the Eastern Interior region, down from 73.1 percent in 1918.[17]

Threatened with destruction of the union, Lewis finally permitted wage reductions in northern union mines beginning in 1927. By 1932 the average daily rate in union mines had fallen to $4.58 although by then the nonunion wage rate had declined to $2.50 for a nine- to ten-hour work day.[18] But out of this chaos came a return to the *status quo ante* following passage of the National Industrial Recovery Act, which touched off the greatest unionizing drive in labor's history. Within months of the Act's passage the U.M.W.A.'s membership rose from 100,000 to 400,000. The first Appalachian Wage Agreement signed in October, 1933, covering most of the industry, established a rational wage pattern, though the south still maintained a 40 cents per day differential under northern mines.

It is against this background of intense wage competition in the 1920's and early 1930's that we view the pressures for mechanization building up in the midwest. Weakened by southern coal competition, Illinois and Indiana operators saw in technological improvement a means of regaining their former market position. They began a technological revolution which continued long after the original need for it had disappeared. It is noteworthy that this

[16] Wilbert G. Fritz and Theodore A. Veenstra, *Regional Shifts in the Bituminous Coal Industry*, Bureau of Business Research, University of Pittsburgh, 1935, p. 61. State consumption figures exclude locomotive and vessel fuel.
[17] Harline, p. 94.
[18] *Ibid.*, p. 85.

revolution preceded by two decades the phenomenal post-World War II increase in productivity witnessed in the rest of the industry.

Technological change took a number of forms, the most notable of which were increases in the percentage of coal mechanically cleaned and mechanically loaded. Tables 25 and 26 show the results of these developments.

TABLE 25. Percentage of bituminous coal production mechanically cleaned

Region	1930	1935	1940	1945	1950	1955	1959
Indiana	1.6	12.3	27.0	44.2	70.7	70.1	70.2
Illinois	N.A.	11.1	37.2	42.1	84.6	84.6	95.5
United States	8.3	12.3	20.3	25.6	58.7	58.7	65.5

N.A. = Not available.
N.B. Western Kentucky figures unavailable.
Source: Bureau of Mines, *Minerals Yearbook*, various years.

TABLE 26. Percentage of underground bituminous coal production mechanically loaded

Region	1925	1930	1935	1940	1945	1950	1955	1959
Indiana	4.8	32.3	62.5	83.9	90.3	95.8	97.2	97.7
Illinois	1.9	48.0	55.3	78.6	87.3	92.4	98.4	99.2
United States	N.A.	10.5	13.5	35.4	56.1	69.4	84.6	86.0

N.A. = Not available.
N.B. Western Kentucky figures unavailable.
Source: Bureau of Mines, *Minerals Yearbook*, various years.

Upgrading coal by mechanical cleaning permitted midwestern operators to differentiate their product from the raw coal predominantly mined elsewhere in the industry. This development was a delayed reaction to the pressure of southern wage competition, reaching its peak during the 1940's. It improved the midwest's competitive position and permitted it to recapture tonnage lost to southern mines.

The widespread introduction of mechanical loading equipment

in the midwest represented a direct attack on the problem of meeting substandard nonunion wages. In the decade from 1925 to 1935, for example, Indiana increased the percentage of its underground coal loaded mechanically from 4.8 percent to 62.5 percent. The rest of the industry in 1935 continued to load by hand nearly 90 percent of its underground production, a costly process nullifying the gains from having low wage rates. Increased underground mechanization and greater reliance on strip mining in Indiana combined to raise output per man-day from 5.19 tons in 1922 to 9.40 tons in 1938. In District 8, the country's largest coal mining district comprising mines in eastern Kentucky and West Virginia, output per man-day during the same period moved from 5.10 tons to 5.16 tons.[19] The rise in the percentage of Illinois and Indiana coal mechanically loaded continued to increase over the years, and by 1959 the process of conversion from hand to mechanical loading had been nearly completed. But the differential advantage earned by increased mechanization had been vitiated by the tendency elsewhere in the industry belatedly to follow the midwest's lead.

The situation in western Kentucky differed from that in Illinois and Indiana. There, as in much of the rest of the industry, low-wage, nonunion operations predominated in the 1920's and early 1930's, and indeed, continued through the 1940's. Though there are no Bureau of Mines' data for western Kentucky tracing the rate of introduction there of mechanized cleaning equipment, industry observers recognize that the district lagged behind the rest of the midwest in this development. The failure to upgrade its coal through cleaning limited the district's market area. Since World War II, however, many western Kentucky operators have installed mechanical cleaning equipment, increasing consumer acceptance of their products. This development plus the greater postwar mechanization of western Kentucky mines have enhanced their market positions considerably.

All three districts have advanced beyond the balance of the industry in their use of strip mining, a technological development giving them a strong differential cost advantage over eastern mines

[19] Harline, p. 97.

relying mostly on the less productive underground-mining method. The midwest's relatively favorable natural conditions have led to this position of strip-mining supremacy, but the spur of low-wage competition previously mentioned doubtless accelerated the movement.

Significantly, the proportion of production in each state mined by stripping increased as the percentage of coal mechanically cleaned increased. Notice in Table 27, for example, that western

TABLE 27. Percentage of bituminous coal production mined by strip method

Region	1925	1930	1935	1940	1945	1950	1955	1959
Indiana	15.4	34.2	43.1	53.2	53.5	53.8	69.4	68.6
Illinois	2.7	11.6	17.9	25.8	23.2	31.3	40.7	48.1
W. Kentucky	0.0	0.0	0.3	9.8	9.6	47.7	44.7	58.8
United States	3.2	4.3	6.4	9.2	19.0	23.9	24.8	29.3

Source: Bureau of Mines, *Minerals Yearbook*, various years.

Kentucky's big increase in strip mining occurred in the post-World War II period when mechanical loading gained a foothold in the district. The existence generally of more impurities in raw strip-mined coal than in underground coal forces most strip operators to wash their products to insure market acceptability. Thus the price of a productive strip operation is the added investment of costly washing facilities.

Can we assess the structural influence of the foregoing technological developments? Three points stand out. First, the early introduction of cleaning plants in the midwest strengthened a market position weakened by the fierce price competition from wage-cutting southern operators. Increased adoption of washing facilities in the late 1930's and 1940's permitted the midwest to regain lost markets and to capture new ones from competitors outside the midwest. Furthermore, this development provided those midwestern operators mechanically cleaning their coal a differential advantage over other operators in the midwest continuing to ship raw coal.

Second, vigorous expansion of production from strip mines and from mechanized underground mines, by cutting costs, reduced the pressure from outside district competition. And it erected a cost barrier *within* the midwest between the mechanized and non-mechanized mines, giving the former a considerable competitive advantage over the latter.

Finally, there is an association between mine or company size on the one hand and the degree of mechanization on the other suggesting that greater mechanization is a partial determinant of growth. Contributing to this condition is the cost (hence competitive) advantage accruing to mechanized operations permitting their growth at the expense of higher-cost mines. Table 28 shows

TABLE 28.   Relation of capital asset size to sales, bituminous coal industry, 1959

Asset size	Capital assets[a] (less reserves)	Sales[a]	Output/Capital
−$50,000	$ 4,453	$ 42,880	9.7:1
50–99,999	10,450	34,488	3.3:1
100,000–500,000	56,703	163,514	2.9:1
500,000–2.5 million	100,525	193,258	1.8:1
2.5–10 million	135,830	260,875	1.8:1
10–50 million	429,433	576,825	1.3:1
+50 million	464,581	594,642	1.2:1

[a] Thousands.

Source: Internal Revenue Service, *Statistics of Income (Corporations),* 1959–1960, p. 69.

the relation between size and capital intensity using a crude output/capital ratio as an indicator.[20] As a corollary, asset growth changes the structure of costs by reducing direct labor costs at the expense of increased amortization charges. Moreover, the progressively lower ratios for companies in each group up to the largest-asset sized class indicate that economies from mechani-

[20] Table 28 uses capital assets instead of total assets in the denominator of the output/capital ratio since this figure indicates better than total assets investment in plant and equipment. Nevertheless, substituting total assets for capital assets little affects the relations exhibited in the table.

zation continue as companies grow beyond the intermediate size range. More important: they suggest that a concomitant of growth must be sufficient financial strength to underwrite high levels of capital expenditure. This condition reflects on the height of entry barriers, an important structural factor to which we now turn our attention.

CHAPTER VI

Barriers to Entry

The influence of entry-barrier conditions on conduct and performance dictates a study of this important structural characteristic. This analysis is doubly important in view of the prevailing tendency to attribute much of the coal industry's chronic instability to easy entry.

Barriers can be grouped into several categories.[1] First, existing firms possess advantages over potential rivals if they enjoy absolute cost advantages. These can be acquired in a number of ways. Established firms can control low-cost production techniques resulting from patents. Additionally, imperfect factor markets can permit existing companies to buy goods and services at prices below those paid by potential entrants. The possession of favorable resources in an extractive industry can have an enormous impact on production costs, giving the favored companies a significant advantage over potential rivals. (Here the term "costs" is used in the accounting sense. Costs are equalized if rents are imputed to the favorable resources.)

Existing firms may command product differentiation advantages over potential entrants from the use of trademarks and skillfully conceived sales promotion and advertising programs. Or, established firms may control better distributive outlets than are available to potential entrants. This is particularly true if their supply is limited through the use of franchise arrangements.

Other factors impede entry as well. The need for large financial outlays to erect efficient plants can deter entry by barring small producers from the industry. We touched on this point in the

[1] Joe S. Bain, *Barriers to New Competition* (Cambridge: Harvard Univ. Press, 1956), pp. 12–19.

previous chapter. In addition, Bain suggests that the existence of scale economies deters entry if the optimum firm's output is a significant fraction of total industry output.[2] If this condition obtains, the entrance of a firm of optimum size depresses price. If the optimum production rate is not reached, price may be unaffected but the entrant's costs rise. Either alternative is unsatisfactory to the entrant, and this condition presumably raises a barrier to entry, permitting existing firms to raise prices above the competitive level.

There may be another important entry deterrent — one that has been overlooked in the literature. This deterrent is the existence of different subjective considerations of what constitutes a normal profit. There may be a gap between the normal profit which keeps a firm in the industry and an entry-inducing profit which is high enough to overcome the risks of entering the industry. Established firms can exploit this gap by raising prices up to the point at which the resulting supernormal profits of the marginal firm equal the entry-inducing profits of the most favorably situated potential entrant.

What leads to these different subjective profit considerations? Existing firms require lower profits to remain in the industry than outsiders need to induce their entry because the former, merely by surviving, have proved that they can overcome the risks inherent in the industry. The outsiders remain untested. Moreover, existing firms have created their reputations, developed their products, established their distribution connections, whereas the newcomers still have these obstacles to overcome. Even if outsiders are not barred from efficient productive processes by patent barriers nor denied distribution facilities through franchise limitations, they may have two strikes against them simply because they are on the outside looking in.

It might be argued that existing firms and potential entrants are on an even footing provided a free market exists for the various productive factors, and especially if capital is freely available. If these conditions exist outsiders can duplicate going concerns' productive facilities, market positions, etc., by hiring identical factors, including the coordinating arm of management.

[2] *Ibid.*, p. 13.

A gap, nevertheless, may still exist from the possession of superior information by existing firms.

In coal mining the gap between existing firms and potential entrants is probably wider than it is in manufacturing. Like the potential entrant in manufacturing, a newcomer in coal must cope with the risks of enterprise which the existing firm merely by its survival has presumably overcome. But in coal mining (as in most other types of mining) risks for the untested outsider exceed those facing entrants in manufacturing. The coal operator must add to the usual risks the uncertainties involved in the mining process as well. The existing, successful mining firm has proved that it can cope with seam, roof, drainage, overburden ratio, and other conditions that heavily influence costs. The outsider can reduce these uncertainties somewhat by extensive drilling, but there still remains an element of risk missing in most manufacturing operations. For example, no amount of test drilling will uncover an unsatisfactory roof condition which, for its correction, elevates costs above planned levels. In neither underground nor strip mining can drilling remove uncertainty concerning coal thickness and quality. The earth contains these secrets, and though drilling helps to unlock them, the operator finally reaches a point beyond which added drilling costs exceed the gains achieved from the reduction of uncertainty. Since drilling can be extremely costly if extensively carried out, the larger, generally more affluent operators are better equipped to reduce the risk of entry than are their smaller competitors.

In the balance of the chapter we shall look at four specific factors to determine their influence, if any, in limiting competition and in impeding entry into the midwest coal industry. We shall study:

(1) The nature of midwestern coal reserves ownership, distinguishing between potential costs of producing existing strip and deep coal.

(2) The extent to which midwestern coals are differentiated through advertising.

(3) The effect on entry of the trend toward the use of long-term coal contracts.

(4) Developments in energy transportation with entry-barrier implications.

Coal Reserves Ownership

There is no barrier to the establishment of new concerns and the development of new properties. . . . Deposits of bituminous are widely scattered; title to workable seams is distributed among thousands of owners. Much of the supply is so readily accessible that mines can be opened quickly and at small expense. Any person or group who can muster a moderate amount of capital is free to enter the field.[3]

This view, expressed in one of the Temporary National Economic Committee's monographs in the 1930's, accurately reflected conditions as they then existed in the coal industry. It would still apply to much of the industry today. But it would not be valid for the midwest on two counts: capital requirements have grown with increasing mechanization and greater concentration of ownership, and *desirable* reserves are less accessible than they formerly were.[4] The latter phenomenon, too, is an outgrowth of increasing concentration and is due also to the changing fuel requirements of the electric utility market.

The solutions to interesting economic questions hinge on the magnitude and availability of coal reserves. Under given techniques, what are the positions of the industry's long-run cost curves when operations are conducted under different resource conditions? That is, to what extent will diminishing returns increase mining costs as the more desirable seams are worked out and operations move into less favorable situations? What is the degree of concentration of ownership of the higher-grade seams? To what extent do reserves uncontrolled by existing operators offer a threat of entry when industry profit margins increase?

[3] U.S. Temporary National Economic Committee, Monograph no. 21, *Competition and Monopoly in American Industry* (Washington, D.C.: U.S. Government Printing Office), p. 24.

[4] Christenson argues that it does not apply to the industry at large as well. ". . . it is preposterous to say, 'there is no barrier to the entrance of new concerns and the development of new properties.' The barriers are clear and easily identified; they grow out of the character of the geological foundations upon which the industry rests." Christenson, *Economic Redevelopment in Bituminous Coal*, p. 115.

All of these problems bear on entry conditions and reinforce our need to understand the coal reserves picture.

Recoverable coal reserves in the Eastern Interior region in 1953 stood at approximately 100 billion tons. Unmodified, these reserves were sufficiently great to meet the fuel needs not only of the midwest but of the nation as well for hundreds of years. But the real costs of mining various segments of these reserves would vary markedly. Wide variations occur in their quality, location, seam thickness, roof conditions, and overburden depth. All of these factors influence coal reserve values. It would be unrealistic to assume therefore that all midwestern reserves are equally attractive. Some are on the margin of desirability, ready to be exploited when demand increases or old reserves are depleted. Other less desirable reserves, thin and under heavy cover, perhaps will never be mined. An analysis of the Bureau of Mines' reserve classification system clarifies this point. Their estimate of total United States' coal reserves includes not only "measured" reserves but "indicated" and "inferred" reserves as well. In the few states whose reserves are classified by thickness and depth, only 5 percent of the estimated reserves include "measured" seams more than 28 inches thick and at depths of less than 1000 feet. Twenty percent are "indicated" reserves over 28 inches thick, and at depths less than 2000 feet. The remaining 75 percent of the reserves are only "inferred" and include seams as thin as 14 inches and at depths as low as 3000 feet.[5]

These figures become more significant when related to the character of reserves mined today. Unlike the less productive European coal industries, American underground mining is carried out at comparatively shallow depths. None of the midwestern mines, for example, works seams at depths exceeding 1000 feet, and most of them operate at 500 feet or less below ground. Moreover, comparatively little production comes from seams as thin as those included in the reserve estimate. In 1950, only 35 million tons — less than 8 percent of national production — was produced from mines working seams less than 3 feet thick. Less than 3

[5] Paul Averitt, Louise R. Berryhill, and Dorothy A. Taylor, *Coal Resources of the United States,* Geological Survey Circular no. 293, Oct. 1, 1953, p. 13.

percent came from mines working seams 30 inches thick or less.[6] In this context it is obvious that recovering the majority of our reserves can be accomplished only by working seams considered today to be unminable, or at best, economically inferior. Because of seam irregularities, property abandonment, and other reasons, large, contiguous reserves are even less available than the reserve estimates would indicate. In a period when the trend is toward large mines, backed up by sizable reserves, this condition assumes importance.

No data are available to gauge the quantities of reserves most likely to be brought into production. But the scaling device in Table 29 crudely represents the relative value rankings of differ-

TABLE 29. Value scale ranking of midwestern coal reserves

Value ranking	Strip			Underground		
1	A1					
2	A2	B1				
3	A3	B2	C1	A1		
4		B3	C2	A2	B1	
5			C3	A3	B2	C1
6					B3	C2
7						C3

ent midwestern strip and underground reserves. It permits us to analyze the relationships between different coals even though we cannot assign specific tonnage values to the various reserve classes.

This value scale ranks midwestern reserves according to quality and mining costs. Letters denote costs, A coal having the lowest mining costs and C coal the highest. The numbers refer to quality, number one being the best. Thus an A1 classification represents coal of the highest quality, capable of being produced at the lowest cost. C3 coal stands at the other end of the cost-

[6] W. H. Young and R. L. Anderson, *Thickness of Bituminous Coal and Lignite Seams at all Mines and Thickness of Overburden at Strip Mines in the United States, in 1950,* Bureau of Mines, Information Circular no. 7642.

quality scale. All strip reserves are classified first with reference to other strip reserves. Then each classification is correlated with the different underground groups which also are internally classified.

How can this scale be interpreted? It indicates that, aside from locational advantages which certain reserves possess over others, the low-cost, high-quality strip reserves will enter the industry before any other reserves. If A1 strip reserves are marginal in the sense that they will be the next reserves called into use, all other strip and underground reserves are submarginal. Moreover, existing operations mining A1 strip reserves presumably earn the largest profits in the industry and are best able to withstand the vagaries of the market. When others around them are falling, they will still stand.

Moderately low-cost, high-quality strip reserves (B1) or low-cost, medium-quality strip acreage (A2) will be developed before even the best underground reserves are exploited. Only at the third ranking tier where A3, B2, and C1 strip coal is developed will the comparatively low-cost, high-quality underground coal be called into production.

The scale considers only quality and production costs, ignoring locational factors which, in view of the importance of transportation charges, can influence the choice of reserves brought into production. This adds a third dimension to the scale which makes it unwieldy. Locational effects, however, can be handled in one of two ways. Either costs can be adjusted to include transportation charges so that the letters denote delivered costs, or favorably situated reserves can be moved up the scale. For example, A1 underground reserves located on or near a navigable stream might move up from the third to the second or first level.

Do the additions to midwestern mine capacity in the past few years bear out the validity of this scale? The answer is that they apparently do. Most of the large new mines opened by Peabody Coal Company and Ayrshire Collieries in the recent past, *which have represented net additions to capacity,* have used A1 to A3 strip reserves. The lower-quality reserves are favorably located, which move them up the scale. United Electric's Banner Mine probably uses B1 reserves but they too are advantageously situated adjacent to the Illinois River. Pittsburg and Midway Coal Company and West Kentucky Coal Company have successfully developed A1 underground mines alongside the Ohio River. Few reserves of coal similarly situated remain unexploited in the midwest. Almost all A1 underground reserves developed in recent years that have not had particular locational advantages have replaced depleted reserves and have not represented net additions to capacity. For this reason they could not have been considered as having "entered" the industry. They could succeed in preference to other reserves on the same or higher value levels because of the gap between their proved profit position and the profit uncertainty for outsiders.

The only recent underground development which has not represented net addition to industry capacity and which uses reserves lower than the third level is Ayrshire Collieries' new Thunderbird mine. It supplies 100 percent of Indiana and Michigan Electric Company's fuel requirements at its new generating station in western Indiana, a few miles from the mine. Significantly, this mine which apparently partially invalidates the value scale is a failure. Bound by a fifteen-year fuel supply contract, the mine's costs far exceed the management's estimates, dooming the property to continual losses or minimal profits for the life of the contract. Ayrshire evidently felt the reserves ranked higher on the scale than they in fact were.

A superficial review of other existing mining operations apparently invalidates the scaling device as well. Yet how can one account for the simultaneous existence of A1 strip mines and B2 underground mines when coal higher on the scale remains unexploited? Several factors contribute to this condition. First is the modifying influence of locational forces already discussed. More important is the ability of marginal and submarginal firms to exist in an industry long after profits have declined below a normal level. Older mines, begun when demand temporarily increased, hang on in the face of adversity by "living off their depreciation." Failing to provide for capital replacement, they eventually die but their demise can be agonizingly slow. In addition, imperfect product markets permit sufficiently distorted pricing patterns to let the established firms often sell at higher prices than the quality of their coal would seem to allow. Finally, the previously mentioned gap between established firms' profit margins and entry-inducing profit margins permits distortions.

Always heavily dependent on stripping, the midwest has turned increasingly to this mining technique in recent years. In 1940, strip mines accounted for 30.7 percent of midwest tonnage; by 1959 this figure had risen to 54.5 percent, in contrast to 21.7 percent for states outside the midwest. This shift in emphasis from underground to strip mining bears out the inherent cost advantage of the latter method. Table 30 measures the extent of this advantage. The mines included in this table represent 89 percent of Indiana's 1952 production. Costs correspond to the accountant's, not the economist's, concepts. Thus there are no imputations to certain factors; rather, all costs are recorded, bookkeeping costs. Each firm belonging to the Coal Trade Association of Indiana reported monthly costs for each mine to the Association. The use of a standardized form insured a uniform reporting of costs. Data were compiled by the Association and issued to its members on a confidential basis.[7]

Note that the pattern of costs corresponds roughly to the one hypothesized in the value scale model. The highest-cost mines are all underground operations, the lowest-cost, strip. Toward the middle of the distribution, high-cost strip mines overlap with the lowest-cost underground mines. There is no reason to believe that this relationship between Indiana underground- and strip-mine costs does not apply equally well to Illinois and western Kentucky mines.

What characteristics determine the mining cost rating for reserves on the value ranking scale? To justify using large, cost-reducing capital equipment, a strip operation must possess a large, contiguous block of low ratio reserves.[8] Unless the reserves are substantial, exorbitant depreciation charges, resulting from

---

[7] When the Association disbanded in 1959 it turned over all of its distribution and cost records to Indiana University's rare book library from which the author secured the data in Table 30.

[8] "Ratio" here refers to the relationship between coal thickness and overburden depth. It can be determined in one of two ways: either as the ratio of seam thickness to overburden thickness or of recoverable coal tons to cubic yards of overburden moved. Both methods give approximately the same results. The higher the ratio, the higher costs will be, other things remaining the same.

TABLE 30. Frequency distribution of average costs, Indiana mines, 1952

Cost bracket	Mechanically loaded shaft mines (no. of mines)	Strip mines (no. of mines)
Over $5.00	1	0
4.91–5.00	0	0
4.81–4.90	0	0
4.71–4.80	0	0
4.61–4.70	1	0
4.51–4.60	1	0
4.41–4.50	1	0
4.31–4.40	3	0
4.25–4.30	2	0
4.21–4.24	0	0
4.11–4.20	1	2
4.01–4.10	1	1
3.91–4.00	0	2
3.81–3.90	0	4
3.71–3.80	0	1
3.61–3.70	1	1
3.51–3.60	0	1
3.47–3.50	0	3
3.41–3.46	0	2
3.31–3.40	0	1
3.21–3.30	0	0
3.11–3.20	0	3
3.01–3.10	0	0
Less than 3.00	0	3
Total	12	24

Source: Materials furnished by courtesy Indiana Coal Association through Lilly Library, Indiana University.

the high initial investment, make mining costs unattractive. If smaller, less costly equipment is used, however, direct mining costs rise. Neither method of operating on inadequate reserves can result in average costs low enough to match the costs derived from a large-scale operation using large reserves.

The most advantageously situated underground reserves are the thick seams under comparatively shallow cover, amenable to mechanized mining operations which are accessible to cheap water transportation and which possess high heat value. These reserves are pretty well confined to the southern Illinois no. 6 and western Kentucky no. 9 coals.

The significant question, therefore, from the entry-barrier standpoint is: who controls the most desirable reserves? The seven companies whose stocks are publicly held and for whom coal reserve data are available, controlled slightly less than 5 billion tons of unmined coal in 1960.[9] This total represented an approximately 100-year supply of coal based on the companies' 1960 rate of operations. Assuming the rest of the midwest's largest producers possess proportionally fewer reserves, it is questionable whether the top twenty companies control more than 7 billion tons, or only 7 percent of all midwestern reserves. Important for our discussion, however, is the fact that virtually all of the *most desirable* reserves are controlled by these comparatively few companies.

The control is probably tightest with strip reserves. A handful of large strip-mine operators control virtually all of the large strip reserves which are capable of supporting annual outputs exceeding a half-million tons. Recent increases in the sizes of stripping equipment have made available for stripping large acreages of coal land heretofore not amenable to the strip-mining technique. But these reserves are adjacent to properties now being worked and are, therefore, easily controlled by existing operators.

The only remaining available strip reserves are small deposits incapable of supporting large, efficient operations. Some of these marginal properties have been developed in southern Illinois and western Kentucky in recent years to supply a part of T.V.A.'s coal requirements. These mines exist only by paying substandard non-union wages. For the most part they are weakly financed and, possessing scanty reserves, they remain in the industry for comparatively short periods of time.

Ownership of underground reserves is more widely diffused but since deep mining costs generally exceed strip costs, most of these reserves pose no immediate threat of entry. The most desirable underground reserves — those standing at the top of the

---

[9] Tonnages were secured from the corporate reports included in Moody's *Industrial Manual,* 1961, and Standard and Poor's *Corporation Records,* 1961.

value ranking ladder — are controlled by a few existing midwestern companies. Thus when Consolidation Coal Company, a large eastern producer, recently expanded its reserve holdings into the midwest, it had to be content with acquiring potentially low-cost but marginal-quality coal in central Illinois. The most desirable reserves were already tied up by midwestern operators.

The only factor capable of affecting the relative desirability of different coal reserves would be a dramatic change in the relative cost of mining strip and underground coal. A large cost-reducing innovation in underground mining would perforce raise the value of all deep reserves relative to strip reserves, shifting upward each tier on the right side of the value ranking scale. If they shifted sufficiently to give underground mining an absolute cost advantage over strip, vast reserves of deep coal, now submarginal, would pose a continuing threat of entry. Though the gap between underground and strip costs may continue to narrow, there is no immediate prospect of strip mining losing its absolute advantage. *Coal Age*, the leading coal trade journal, predicts that between 1960 and 1970 productivity gains in deep mining will run 26 percent ahead of those achieved in strip mining; yet by the later date average strip productivity will still be 65 percent ahead of deep, measured in output per man-day.[10]

Is it not possible for entrants to overcome the entry barrier imposed by the unavailability of the "best" reserves by securing less attractive reserves at a lower price or royalty rate? If the value ranking scale we have crudely constructed is valid, then prices of coal reserves lower on the scale should reflect their undesirability. If the differentials between prices of the most favorably situated reserves now under control and less desirable reserves reflect each reserve's scarcity value, entrants should be able to surmount the reserve ownership entry barrier by procuring "poorer" reserves. The character of the market for coal reserves (discussed in detail in Chapter VIII) prevents this development. Cost and quality advantages associated with the better reserves give operators mining these deposits an enormous edge over would-be entrants forced to use second-rate reserves. Even at

[10] "A Half Century in Coal and the Next Ten Years," *Coal Age,* Oct., 1961, p. 130.

a zero price or royalty rate it would be infeasible to exploit many reserves; the great bulk of the reserves would require a negative price to call them into production.[11] Of course the relative ranking of coal reserves is a dynamic process. Reserves which today could not be purchased except at zero prices, may achieve value as more desirable reserves are worked out. Thus, even in the absence of technological developments upgrading underground coal, the rapid depletion of prime strip reserves sharply enhances the potential value of remaining deep reserves.

The increasing concentration of the most desirable reserves has had an impact in the market place. Witness the effect on T.V.A.'s coal bids. In the three and a half years between January, 1958, and July, 1961, the average delivered coal price on T.V.A.'s coal bids for plants using midwestern coal rose while the number of bids steadily declined. Twenty-eight Illinois and western Kentucky operators submitted bids for the January, 1958, bid opening for the Shawnee plant at delivered prices averaging 16.65 cents per million B.t.u.[12] T.V.A. received only eleven bids from ten companies at the July, 1961, Shawnee bid opening, the average price being 17.63 cents per million B.t.u., 6 percent higher than before. Six of the ten companies represented were among the eight largest midwestern producers. The same pattern prevailed on bids to the Johnsonville steam plant, another T.V.A. plant in the midwest's market area. Between the two periods cited above, the number of bids declined from twenty to eight (from seven companies); the average mine price level increased 7 percent. In both cases large companies dominated the bidding, most of the small suppliers having been squeezed from the market.[13]

Contrast this trend with the bidding performance of eastern

[11] Putting a negative value on some coal is not as fatuous as it first appears to be since there are "holding" charges (taxes, drilling, etc.) that detract from a reserve's value.

[12] These and succeeding figures were derived from T.V.A. bid sheets generously furnished by Mr. E. C. Hill, Chief of Coal Procurement, T.V.A.

[13] It would be erroneous to conclude that, this being true, midwestern shippers possess a degree of monopoly power in the T.V.A. market permitting them to raise prices above the competitive level. There is still vigorous price competition in this market, more perhaps than in any other major segment of the midwestern industry's market. The significant point for the present discussion is the trend toward decreasing competition that has accompanied the increasing concentration in reserve ownership.

operators vying for sales to T.V.A.'s Kingston plant in eastern Tennessee. At the same 1958 bid opening twenty-six companies furnished bids at an average delivered price of 20.52 cents per million B.t.u. Three and a half years later there were still twenty-four bids and the average delivered price level had fallen to 19.92 cents. The delivered price per ton unadjusted for B.t.u. content, dropped 40 cents, or over 8 percent.

Effect of Product Differentiation on Entry

Chapter III indicated that coal is far from being a homogeneous commodity. Coals differ widely from each other in terms of their chemical constituents. But the narrower the coal region studied the smaller are these variations. Midwestern coal operators have tried, however, to widen the differences artificially through improved beneficiation and, moderately, through advertising. Neither approach has given established operators significant long-run marketing advantages over competitors; they have been even less effective in impeding entry.

As indicated in Chapter V, the midwest has led the rest of the industry in establishing washeries to improve coal quality. This development was a child of necessity. Reeling from the competitive pressure of cheap nonunion coal in the late 1920's, the midwest sought relief through the use of two measures: the installation of mechanized mining machinery aimed at cutting costs, and the increased use of washing plants to improve midwestern coal's acceptability. The latter development gave those operators who installed washing plants a competitive edge over raw coal producers. From the long-range standpoint, though, the installation of washeries is a weak weapon in the product differentiation arsenal. There is nothing to prevent competitors from building duplicate washing facilities, nullifying the initiator's advantage. It is an equally weak device for discouraging potential entrants since they too can erect identical facilities.

Midwestern operators are just as ineffective in securing product differentiation advantages through advertising. This is not surprising. Few, if any, relatively homogeneous, industrial bulk commodities lend themselves to effective promotion, unlike consumer products, whose producers often can erect formidable entry barriers through the use of massive advertising expenditures. Recognizing this difficulty, midwestern producers have done little to manipulate demand for their product through advertising. Ten large midwestern operators accounting for 69 percent of the midwest's 1960 production, reported spending in that year only $486,536 on advertising, or less than seven and one-half mills per ton.[14] Asked what specific benefits the company felt it received from advertising, most respondents who were mailed questionnaires were hard pressed to justify the expenditures. Clearly, advertising offers little opportunity for deterring entry into this industry.

Trend Toward Long-Term Coal Contracts

The rapid growth of coal consumption by electric utilities, together with their dependence on reliable long-range fuel supplies have led to the increasing use of long-term coal contracts. "Long-term" in this context refers to contracts with a duration of more than one year. In most cases they run for from three to fifteen years, and in some instances, as long as thirty years. Table 31

TABLE 31.   Distribution of 1960 midwestern production by contract length (tonnages in thousands)

Company size (annual production)	Spot (tons)[a]	Per-cent	One-year (tons)[b]	Per-cent	Long-term (tons)[c]	Per-cent	Total (tons)	Per-cent
+1 million	7,432	12.1	9,560	15.6	44,206	72.3	61,198	100
250,000–1 million	1,009	54.8	625	33.9	207	11.3	1,841	100
−250,000	190	40.2	219	46.7	62	13.1	471	100

[a] Tonnage sold on open order or on contracts with a duration of less than one year.
[b] Tonnage sold on contract with a duration of one year.
[c] Tonnage sold on contract with a duration of more than one year.
Source: Mailed questionnaire.

indicates the widespread use of long-term contracts, especially among the largest producers. In 1960, midwestern operators producing over one million tons a year who responded to a questionnaire sold 72.3 percent of their output under long-term contracts.[15]

[14] This figure was derived from a questionnaire mailed to eighty midwestern coal companies.
[15] These producers accounted for 79.9 percent of the coal produced by companies in this size group.

Ten years previously the same companies, then producing 35.8 million tons, sold only 10.4 million tons, or 29.1 percent of their production on a long-term basis. In 1950 only two of the eleven reporting firms shipped more than half of their coal under long-term contracts; five shipped from 10 to 30 percent on this basis and four shipped none. In contrast, by 1960, no producer shipped less than 20 percent on long-term contracts, eight of the eleven selling more than 50 percent under long-term commitments.

The ramifications of this development are widespread. First, it affects different-sized producers unevenly. Notice in Table 31 that the small- and medium-sized operators mining less than a million tons annually shipped only 13.1 percent and 11.3 percent, respectively, on long-term contracts, concentrating more on short-term sales. This condition could have arisen from a preference of the smaller producers to specialize in short-term sales. This explanation is partly valid, especially for those producers catering to the retail trade. But for many, long-term contract sales were small because their coal reserve position and sales connections were inadequate. Nine of the ten respondents to the questionnaire who produced less than 500,000 tons annually thought the larger producers possessed sales advantages that they, the smaller producers, lacked. Asked what advantages the large producers had which prevented small operators from competing effectively, they cited: "better reserves," "better sales contacts," and "able to get big contracts."

To qualify for most of the long-term contracts now being negotiated, a firm must control a sufficiently large block of desirable reserves to guarantee delivery over the life of the contract. These reserves requirements may appear overwhelming to small producers. For example, Consolidation Coal Company's contract with Central Illinois Public Service Company requires the producer to dedicate 125 million tons of reserves for the power plant's use over the life of the contract.[16] For every coal reserve

[16] In the period prior to World War II mines were smaller and reserves requirements less than they have been in recent years. Rarely did a mine begin with its output presold as is true today. Then, an operator might sell 20 to 25 percent of a new mine's output to the railroad serving the mine as locomotive fuel and attempt to sell the rest of it on the open market using price-cutting tactics. The destabilizing influence of the practice sharply contrasts with the relative tranquility existing in today's markets.

of this magnitude brought into production the operator ordinarily has a number of others for which he can find no market outlet. The acquisition of these reserves is costly and represents a sterile investment until mining operations begin. Only large, well-financed firms can support the reserves necessary to compete in the growing utility market in which the long-term contracts are predominantly used.

The second significant aspect of the trend toward the use of long-term contracts is its effect in shifting production from the everyday market place and virtually transforming suppliers under some long-term contracts into captive producers. The extent to which this phenomenon exists depends on the length, size, and type of contract entered into. Any contract — even one expiring in a year or less — reduces price competition if it provides for no price adjustments during the term of the contract. The coal covered by the agreement in effect is removed from the market. Longer contracts remove coal from the market for greater time periods.

The ultimate in this development is the creation of new mine properties whose output is dedicated exclusively to a consuming plant for the life of the coal reserve. Less extreme is the case of the company which contracts to ship most of the mine output for an extended period of time. In each of the latter two cases new mines are usually created to supply coal under the long-term contracts.

Table 32 lists the coal contracts responsible for new midwestern mine developments from 1955 to mid-1963. This list is believed to be complete. It represents information distilled from many well-informed people in the industry. It excludes the many smaller contracts that make up the high percentage of long-term contracts included in the data in Table 31, highlighting, instead, just the contracts that have made new mines virtually into quasi-captive operations.

The third effect of the trend toward increasing use of long-term contracts is a corollary of the second. Removing sales from the vicissitudes of the everyday market, prices are more stable and profits more secure than they are when output faces the

TABLE 32.   Long-term coal contracts responsible for midwestern mine openings, 1955–1963

Company	Mine	Location	Coal consuming plant	Annual tonnage	Term of contract (years)
Peabody	Name unknown[a]	Indiana	Alcoa	1,000,000	30
Peabody	Paradise	W. Ky.	T.V.A.	3,941,000[b]	17
Peabody	River Queen	W. Ky.	Ind. & Mich. Elec.	1,000,000	15
Peabody	River Queen	W. Ky.	Cincinnati Gas & Elec.	1,000,000	15
Peabody	Homestead[c]	W. Ky.	Ind. & Mich. Elec.	1,000,000	10 (?)
Peabody	Victoria	Indiana	O.V.E.C.		15
Peabody–Ayrshire	Gibraltar	W. Ky.	O.V.E.C.	4,000,000	15
Ayrshire	Wright	Indiana	O.V.E.C.		15
Ayrshire	Thunderbird	Indiana	Ind. & Mich. Elec.	1,300,000	15
Midland–Electric[d]	Viking	Indiana	Pub. Serv. Co. of Ind.	700,000	10
Midland–Electric[d]	Green Valley	Indiana	Pub. Serv. Co. of Ind.	800,000	10
Midland–Electric	Captain[e]	Illinois	Commonwealth Edison	4,000,000	10
Freeman	Farmersville	Illinois	Commonwealth Edison	1,500,000[f]	16
Pittsburgh & Midway	Paradise	W. Ky.	Pub. Serv. Co. of Ind.	800,000	12
Pittsburgh & Midway	Paradise	W. Ky.	Kentucky Utilities	300,000	10
Pittsburgh & Midway	DeKoven	W. Ky.	City of Memphis	480,000	10
Pittsburgh & Midway	DeKoven	W. Ky.	Electric Energy, Inc.	500,000	12
Old Ben	Under development	Illinois	Union Electric	1,500,000[g]	20
Consolidation	Under development	Illinois	Cent. Ill. Pub. Serv.	1,000,000[g]	30
Total				24,821,000	

[a] Mine name unknown. Though the contract has been awarded to Peabody, a failure of aluminum demand to match expectations had delayed opening the plant, which is scheduled to burn the Peabody coal, through at least mid-1963.

[b] This is an average figure. Contract calls for the delivery of 67 million tons beginning in 1962 and eventually reaching a peak of 4 million tons annually.

[c] Scheduled to begin operations in 1964.

[d] These two contracts belonged originally to Snow Hill Coal Co. which merged into Midland Electric in 1962.

[e] Scheduled to open in 1964. Tonnage shown is maximum, to be reached several years after mine opening.

[f] Original tonnage called for 1 million tons and 10-year life. Later extended additional 6 years a tonnage listed.

[g] Mines scheduled to open in 1966. Figures are projected tonnages.

Source: Interviews with coal industry and utility executives.

continuing pressure of market forces.[17] Typically, a long-term contract specifies a mine price which fluctuates in response only to changes in production costs. The escalation provisions spell out detailed methods for adjusting prices in response to wage-rate increases, changes in supply costs, and general and administrative cost adjustments. Thus, if an operator carefully controls his costs, the contract guarantees relatively constant profit margins for the life of the agreement. Many contracts even provide

[17] The effect of long-term contracts on prices is explained in Chapter VII.

protection to profit margins from the erosion of inflation, granting price adjustments in response to changes in the general price level.

The contracts possess disadvantages as well. The provisions protecting the operator when demand and coal prices in general fall work against him when conditions are reversed. That is, when coal demand increases, pushing up the general coal price level, long-term contract prices remain unaffected, preventing the operator from capitalizing on the favorable short-run profit possibilities. And woe betide the operator who has miscalculated his costs or his coal quality! The contracts listed in Table 32 which run for ten to thirty years were written prior to each mine's opening. Extensive drilling indicated the extent of the coal reserves, their thickness and depth, and core samples indicated quality. But no amount of drilling can eliminate completely the risks inherent in mining. If unexpected mining conditions crop up to plague the operator or if quality fails to measure up to the level indicated in the test drillings, the blessings of a long-term contract can turn into a nightmare. The contract may be a guarantee of unrelieved losses for its full term. At least two of the operations listed in the table labor under this unpleasant handicap.

To appreciate completely the entry barrier significance of long-term contracts, we need to study two other aspects of them. If they are to serve as effective entry barriers they must not be easily canceled at the buyer's whim, nor can they stabilize prices and earnings if the buyer can readily force price reductions. We interviewed coal operators and utility executives to learn industry practices with respect both to forced price reductions and to contract cancellations. In nearly every long-term contract studied, their provisions give buyers no opportunity unilaterally to cancel or to cut prices. More important — the contract provisions are carried out in practice.

This represents a change from conditions prevailing as late as the mid-1950's. Then, contracts were not very explicit concerning price adjustments to be made for wage and supply cost changes. The late 1940's and early 1950's were marked by frequent changes in these major cost components. Changes in the U.M.W.A.'s wage scale particularly gave sellers and buyers frequent opportunities to negotiate price adjustments. The amount of increase was a function primarily of the two parties' relative bargaining power which was conditioned by the general level of coal demand existing at the time of the negotiations. The bargaining strength of utility buyers took two forms: (1) When wage and supply cost changes required price adjustments, the resulting price increases fell short of the cost increases. (2) When the market softened, utility buyers demanded and usually received price reductions.

Today these conditions no longer exist under the operation of most long-term contracts. Operators complain that occasionally a buyer will bring pressure to reduce prices, but these instances are rare. Also, a few long-term contracts are rigid with respect only to tonnage, permitting negotiations on price once each year. In one contract of this type, failure to agree on price requires the parties to resort to a modified "most favored nation" provision, the resulting price being an average of the producer's other utility prices in the same market area. But selling other utilities on a fixed-price basis effectively stabilizes the price to this utility, notwithstanding the flexibility apparently provided by the contract. Some other contracts insure stable prices but allow reduced tonnages, especially to take advantage of alternate fuel availability. Most of the contracts guarantee the operator a minimum tonnage that adequately protects his position.

The tendency of long-term contracts to foreclose competition from large segments of the market for extended periods of time raises an interesting legal question. Aren't these contracts assailable under the anti-trust laws? The Supreme Court ruled in the Tampa Electric Company vs. Nashville Coal Company decision that the contract was legal under the circumstances in that case.[18]

Nashville Coal Company agreed to supply Tampa's Gannon Station its entire coal requirements for a period of twenty years, the contract tonnage eventually to total 2,250,000 tons annually. Later, seeking release from the contract, Nashville argued before the lower courts that the contract violated the anti-trust laws. In its decision a Court of Appeals declared that the contract

[18] 365 U.S. 320 (1961).

violated Section 3 of the Clayton Act which makes it unlawful to make a sale on the understanding that the purchaser will not use the goods of a competitor, where the effect of such sale may be substantially to lessen competition or tend to create a monopoly. Though the contract contained no conditions expressly forbidding Tampa from buying competitors' coals the total requirements proviso created the same effect.

The Supreme Court reversed the lower court's decision. Its ruling hinged on a definition of the relevant market area where the seller operated, declaring that the "contract must be found to constitute a substantial share of (that) relevant market." [19] The Supreme Court broadened the concept of the market to include the market area covered by the 700 producers in eight midwestern and Appalachian area states capable of serving the Florida market. Buyers in the relevant market area, thus viewed, consumed 250,000,000 tons of coal annually. Tampa's 2,250,000-ton annual requirement, therefore, represented an insubstantial portion of the total market; hence the Court declared the contract legal.

Conditions in the Tampa Electric case are similar to those existing under the requirements contracts between midwestern coal producers and electric utilities. As long as the Supreme Court views the relevant market as broadly as it did in the Tampa case, it is hard to imagine the midwestern long-term contracts coming under attack. There, for example, it included, as potential suppliers, producers from states which have never shipped coal to Florida and which under ordinary circumstances probably never would service that state. If the midwestern contracts were attacked and the Court held that eastern operators can effectively serve the midwestern utility market then each contract would represent an insubstantial share of a total market accessible to hundreds of operators. But should the Court view the midwest as a sheltered market relatively free from outside competition, it might move from its position in the Tampa case.

Whatever the legal arguments may be, there still remain important social policy implications of long-term contracts. The Tampa case to the contrary notwithstanding, a requirements con-

[19] 365 U.S. 328 (1961).

tract effectively bars competition from the segment of the total market represented by *that* contract. Repeat these contracts many times over, and the result is a congeries of protected submarkets "monopolized" by individual operators. And the list of the firms eligible to serve these submarkets grows shorter as the merger movement continues.

Though the Supreme Court declared that "it may well be that in the context of anti-trust legislation protracted requirements contracts are suspect," [20] still it recognized as a mitigating factor the need of utilities to assure themselves long-term, dependable fuel supplies. Another factor modifying the possible disadvantages of requirements contracts is their influence in maintaining a high rate of resource utilization. Being tied to an assured, seasonally stable market should permit mines to make fuller use of capital equipment and labor resources than is possible under usual coal market conditions.

Energy Transportation Developments

Recent developments in energy transportation have coincided with the development of long-term coal contracts. Their entry-barrier ramifications may surpass in importance those flowing from long-term contracts and from the concentration of desirable coal reserves into fewer hands. Indeed, all of these forces are so closely intertwined that it is difficult to tell where one leaves off and the other begins.

The importance of freight charges in coal's delivered price has led to continued efforts to reduce coal transportation rates. Barge movements of coal have provided relief, but only for shippers and receivers having access to navigable rivers. More recently coal pipelines have posed a new threat to the railroads, traditional carriers of most coal shipments. Consolidation Coal Company has proved the coal pipeline technically feasible by operating one for several years between Cadiz, Ohio, and a Cleveland Electric Illuminating Company plant on the Lake Erie waterfront.

Reaction by the railroads to the dagger poised at their throats has been resolute. They have begun a massive restructuring of freight rates to blunt the pipelines' attack. In Ohio the rates have

[20] 365 U.S. 333 (1961).

been reduced low enough to force the pipeline's abandonment. And they have started to fashion plans for totally new concepts in bulk commodity movement. The ultimate in these developments to date is the unit or integral train composed of specially designed cars capable of carrying 100 tons or more each in groups of 100 to 250 cars. The entire train operates as a unit, being broken up for maintenance only. It stops only for crew changes and for rapid loading and unloading at each terminus.

The conditions under which these trains operate permit drastically reduced rates for volume shipments. Already the New York Central Railroad has published a rate of $1.45 per ton for integral train shipments from Lynnville in southern Indiana to Commonwealth Edison's Stateline plant at Hammond, Indiana. The new rate compares with a single-car rate of $3.42 per ton. The Gulf, Mobile, and Ohio Railroad has published a volume rate of $1.30 for integral train shipments from a new mine at Percy, Illinois, to Plaines, Illinois. Doubtless other rates will follow.

These rates have several features in common. The shipper must tender the cars to the railroad *on a single bill of lading, on one day,* and must be *destined to a single consignee.* The rates are subject to minimum-weight restrictions (approximately 10,000 tons). Cars must be loaded and unloaded within four hours of placement to avoid stiff penalties. Either the shipper or consignee provides and maintains the train's cars. Finally, the rates apply only if annual shipments reach a specified level. The applicable minimum for the Percy–Plaines' rate, for example, is 3,300,000 tons for each consecutive twelve-month period.

These conditions all add up to one thing: the granting of a huge competitive advantage to the very large coal operations capable of satisfying utilities' multimillion-ton annual fuel requirements. It puts small- and medium-sized producers alike at a disadvantage. Firms which are incapable of meeting the integral train requirements are in danger of losing existing tonnage in addition to being foreclosed from future utility contracts. The only relief available to the small producer, unable to muster enough tonnage to meet the integral train tonnage requirements, is to pool tonnage with similarly situated operators. Some of the unit train rates in the east permit this practice. There, rates apply to all mines operating within a given district. But such pooling reduces the unit train's inherent advantages. Rates under this arrangement, therefore, are much higher than they are when only one shipper is involved.

There is another piece in the energy transportation puzzle which is needed to complete the picture. It is based on the creation of electric transmission systems capable of carrying increasingly higher voltages over longer distances, an innovation with portentous consequences for several industries, including coal.

Heretofore, most steam-powered generating stations have located near load centers, the coal required to fire the boilers being transported from distant coal fields. What has dictated this locational arrangement? Why haven't power plants been located in the coal fields, with the electricity itself rather than the energy source which creates it being shipped to the point of consumption? The answers have rested with the relative economics of moving power and coal. The high cost of transporting power over long distances has generally ruled out this alternative. Transmission losses accompanying the use of relatively low voltages have prohibited the movement of power over long distances. Thus, the development of higher voltages raises interesting questions concerning the possibility of moving generating stations away from load centers and toward the coal fields.

The trend toward higher voltage is unmistakable. In 1892 power moved over 10,000-volt transmission lines. Today 345,000-volt lines are in use, permitting the location of power plants at or near coal fields. Within the next decade we may witness a milestone in power transmission as the entire continental United States becomes interconnected in a far-flung transmission grid that will permit, theoretically at least, the consumption in New York of power produced in California.

It is difficult to compute the relative costs of the two alternatives — moving coal to the power-consuming point, and transmitting power long distances via high-voltage lines. Too many variables muddy the waters. Table 33, however, throws some light on the issue by translating transmission costs into costs per million B.t.u.

TABLE 33.   Transmission costs in relation to line length

Miles	Fuel cost equivalent (cents per million B.t.u.)	
	500 kv	700 kv
300	11.7	10.0
400	15.2	12.9
500	18.7	15.8
600	22.2	18.7
700	25.7	21.6

Source: J. K. Dillard and E. W. DuBois, "Developments and Prospects in High Voltage Power Transmission," a speech before the American Transportation Research Forum, Pittsburgh, Pennsylvania, Dec. 27, 1962.

We need to compare these figures with the costs of moving fuel by rail or pipeline. A Department of Interior study estimates the costs of moving coal by pipeline from southern Illinois to Chicago, a distance of around 275 miles, to be approximately 10.5 cents per million B.t.u.[21] This figure is competitive with the cost of moving the same energy by wire over 500 kilovolt lines and is slightly higher than transmission costs using 700 KV lines, assuming a 90 percent load factor. It also would be slightly higher than an integral train rate. Implicit in these rough calculations is the assumption that the consuming points are capable of absorbing the huge blocks of power required to justify building large transmission systems and pipelines capable of moving 9 million tons of coal yearly.

Proponents of high-voltage transmission argue that there are collateral advantages to its use more important in the over-all economic evaluation than possible fuel transport savings. First, by interchanging power, the most economical power generating stations can be used, increasing the industry's efficiency rate. Any utility tied into the system can tap the power generated by the mine-mouth power plants that form the base of the system. Power will move freely within the system, with shifting load patterns and generating economies dictating which power plants

[21] Department of Interior, *Report to the Panel on Civilian Technology on Coal Slurry Pipelines,* May 1, 1962, following p. II-23.

provide the system's power. Comparatively simple metering and accounting arrangements make administration of the power swaps manageable. Pooling power also permits the construction of larger generating units than would be possible in the absence of such pooling. And since economies of scale flow from the use of larger equipment in this industry, this development adds to economic efficiency.

Economies derive from several other aspects of interconnection. Power pooling reduces the need for maintaining idle or "spinning" reserves. Any power system must have reserve capacity to provide a cushion for seasonal peaks, for growth, and for the downtime required for maintenance and emergency repairs. But a power interchange system reduces this need since each unit stands ready to support the balance of the system during emergency periods. Additionally, when systems span two time zones, reserve power from the area whose industries come to rest in the late afternoon is available to supply the early evening load in the area to the west. The flow can change directions in the early morning hours.

Furthermore, locating power plants away from cities can reduce air pollution, and using higher voltage trims the costs of acquiring rights of way, an important consideration, particularly near metropolitan areas. Moreover, increasing the reliability of electric service through extensive interchange arrangements has national defense implications that cannot be ignored.

The implications for the coal industry of high-voltage power transmission developments and its concomitant, mine-mouth generating stations, are similar to those of most of the other developments previously discussed. To justify added costs from transmission losses utilities must minimize transportation charges. Thus coal must come only from mines adjacent to the power plant. Those mines listed in Table 32 which fuel mine-mouth power plants are sole suppliers to the generating stations they serve. This is the prevailing practice wherever mine-mouth plants are installed. Purchasing coal from several mines adds transportation charges to the fuel bill, defeating the mine-mouth plant concept. The need for large generating stations requires correspondingly large coal operations supported by vast coal reserves. Again, the small operator's size discriminates against him.

Exit Conditions

Supply conditions are affected not only by the height of barriers influencing entry but also by the speed with which firms leave the industry. Before leaving this general area of examination, therefore, we need to study briefly exit conditions in the midwestern industry.

The traditional treatment of this subject views exit as being sticky. Morton Baratz offers three contributing factors:[22] (1) Mining machinery being specialized, there are few alternative uses to which it may be put when operations become unprofitable and the need for shifting resources to new industries arises. (2) Certain development expenses incurred in the opening of a mine cannot be recovered at all except through mining. (3) Historically, demand for coal has fluctuated widely, offering hope to the submarginal operator that a sudden change in the industry's fortunes will turn his losses into profits if he can continue to operate a little longer.

There is evidence that resources may be more mobile than Baratz and other students of the industry have recognized. Table 34 shows the relation between earnings (pre-tax margins) and industry capacity in Indiana for the post-World War II years that bear on this matter. From 1947 to 1951, in response to the relatively lush profits earned in 1947 and 1948, Indiana underground-mining capacity increased from 13.5 to 16.1 million tons. In five of the next six years, underground mines in the aggregate suffered losses; capacity fell sharply to 7.7 million tons, indicating a fairly rapid withdrawal from the industry in response to adversity. In contrast, the entry response was sluggish as a function of the lag between the decision to open an underground mine and the fulfillment of that decision. The "bloom" was off the coal market by the end of 1948; yet capacity continued to rise through 1951 as previously scheduled capacity came into production.

Resources used in strip mining responded more sensitively to changes in demand than did underground resources. Strip-mine pre-tax margins, in 1949, stood at half the 1948 level and the ca-

[22] Morton Baratz, *The Union and the Coal Industry* (New Haven: Yale Univ. Press, 1955), p. 4.

TABLE 34. Relation of operating margins to productive capacity, Indiana mines, 1946–1956

Year	Strip		Underground	
	Margin[a]	Capacity (thousand tons)	Margin[a]	Capacity (thousand tons)
1946	$0.35	14,710	$0.06	13,390
1947	0.66	15,310	0.30	13,500
1948	0.96	15,550	0.40	14,230
1949	0.49	14,160	−0.11	15,020
1950	0.53	13,110	0.03	15,190
1951	0.33	11,070	−0.11	16,090
1952	0.36	11,710	−0.13	12,600
1953	0.40	12,960	−0.24	10,320
1954	0.25	11,410	−0.27	9,520
1955	0.24	12,870	0.08	8,490
1956	0.30	13,290	−0.03	7,750

[a] Pre-tax operating margin per ton.
Source: Materials furnished by courtesy Indiana Coal Association through Lilly Library, Indiana University.

pacity total reflected the change immediately by falling 10 percent. Resources continued to leave the industry during the following two years as margins fell even lower. The character of the strip-mining process and the nature of the equipment used combine to increase the responsiveness of strip capacity to the industry's changing fortunes. Strip operations ordinarily get into production faster than comparable underground mines because they require less development before coal can be extracted. Underground operations must drive long and costly entries, bringing the mine slowly up to capacity operations over an extended period of time. On the other hand, strip mines attain capacity output quickly after the removal of a "box cut" laying bare an initial quantity of raw coal for loading into trucks.[23] Moreover, the shovels, draglines, trucks, and drills used in strip mining are

[23] A "box cut" is the removal of overburden from the first cut in a new strip coal pit.

readily transferred from other industries when coal demand increases, boosting mine capacity quickly. Likewise, being less specialized than underground-mining equipment, strip capital resources invested in small operations readily leave the industry.

A simple, linear regression analysis, though far from conclusive, indicates the more responsive movement of strip-mining resources. A regression of underground capacity on pre-tax profit margins for the period 1946–1956 indicates a weak correlation, with a value for r of only .11. Lagging capacity two years behind margins improves the correlation coefficient to .26, though the relationship is perhaps too weak not to result from chance alone. Averaging margins for three periods and regressing the average, with capacity lagged two years behind the last of the three years, improve the "fit," increasing r to .50. The improvement in correlation in the latter two cases makes sense. Lagging capacities behind operating margins gives resources time to respond to changes in industry conditions, and averaging margins puts entry and exit decisions on a longer-term footing. Firms are less inclined to move from an industry after an unprofitable year than they are if losses persist over a several-year span. Similarly, resources are more apt to shift into a profitable industry the longer profits remain at attractive levels.

In contrast to the weakly responsive movement of resources into underground mining and out of it, strip resources shift rather quickly. A regression of strip capacity on pre-tax profit margins in the Indiana industry, again for the period 1946–1956, produces an r value of .72 without any time lag.

The exit of labor from the industry in response to decreased demand matched the relatively responsive movement of capital. Between 1951 and 1956, when Indiana underground production declined 39 percent and capacity fell 51 percent, employment dropped 65 percent. Part of the proportionately greater reduction in employment resulted from increased productivity. Most of the absolute employment decline from 6,293 to 2,203 employees, however, stemmed from the exit of workers from the industry. Their exit followed logically from the decline in capital resources devoted to the industry.

Before leaving the question of the mobility of resources, we might view the problem briefly from another angle. Despite evidence above that resources are relatively mobile (at least in the Indiana industry), they may still be immobile from the traditional economic view. We have asked only: do resources freely move in response to changing demand conditions? In addition we should know whether resources remain in the industry after their returns fall below the incomes available to those resources in alternative uses. That is, are the opportunity costs of remaining in the industry greater than rewards earned by continuing to operate? A casual reading of Table 34 indicates that under this criterion, underground-mine capital resources and quite possibly strip resources as well, are immobile.

CHAPTER VII

Market Conduct in the Midwestern Coal Industry

We turn our attention now to market conduct in the mid-western coal industry. We are concerned with the price calculation procedure used by individual firms, with the mechanism by which sellers coordinate their price policies, and with the pricing implications of long-term contracts. Further, we study the influence, if any, of the union in the adjustment of output to meet demand conditions.

Moreover, we view the way in which the product is adjusted for the market. Coal lends itself less readily than other industries to the use of sales promotion and product policies to influence the shape and position of individual firm demand curves. The preceding chapter, for example, indicated the minor role played by advertising as an entry deterrent. Advertising and other sales promotion devices are equally impotent as means of influencing the market conduct of established sellers. The only "product" policy available to coal shippers is the use of different coal sizes which represents an anemic effort to differentiate an essentially homogeneous product. But the conduct of operators individually and jointly in adjusting prices and outputs to market conditions provides the material for more fruitful analysis, and this area receives most of our attention in this chapter.

Interseller Price Relations

Investigators studying the relation of market structure to market conduct often note a tendency toward collusive price determination in industries possessing high seller concentration. In less highly concentrated industries less perfect collusion apparently exists. In view of increasing seller concentration in the midwest, therefore, we should determine whether its pricing conduct falls into this pattern, and if so, the extent and effectiveness of collusion in the industry.

Traditionally, the trade association has served as a structure for organizing collusive price agreements. Until recent years it would have been too difficult to determine which midwestern coal association to suspect of coordinating prices, if indeed such suspicion was justified. Producers in each of the three midwestern states had their own associations organized on a state-wide — and in some instances even on a subdistrict — basis. Most of the associations conducted specialized activities, individual organizations being set up to handle marketing, labor, statistical, and traffic problems of mutual interest to association members. In some instances an association served as a militant arm of operators in a subdistrict waging battles against competing districts. Thus Indiana operators divided into two warring groups seeking to gain freight-rate advantages over one another on shipments into common markets. Illinois operators in different subdistricts created their own traffic associations as well, though competition among them was less spirited than it was in Indiana. To protect their freight differentials western Kentucky producers also operated a traffic association. Occasionally two or more competing associations buried the hatchet long enough to make common cause against a competing district when their interests coincided. The end of the freight-rate case, however, usually terminated the coalition. The normal pattern called for the traffic associations to extend into freight-rate determination the often bitter mine-price competition existing among competing producers.

The suspicion among operators carried over to the labor field, each district establishing its own labor associations on the grounds that special conditions within each district dictated that they handle their own affairs. Indiana again divided into two camps, setting up separate strip and underground associations. A fear that one group would make concessions to the union detrimental to the interests of the other lead to the schism.

Statistical and marketing (or general trade association) activities generally were interrelated. Indiana, Fulton-Peoria, southern Illinois and western Kentucky operators collected sales (and in some cases, cost) data that supplemented their general association functions. Indiana operators, for example, received monthly tabulations presenting average sales realization figures by size group, destination, and seam classification. These reports usually failed to reveal competitors' exact prices to specific customers, but they showed the general level of prices more accurately than salesmen's impressionistic reports indicated. Operators rarely referred to the realization data when determining association policies, but individual operators used them in evaluating their companies' price performance.

More useful as a coordinating device were cost reports submitted monthly to association members who reported their costs on a uniform basis. In Indiana, particularly, they helped operators determine a "fair" price increase to pass on to customers after labor negotiations had raised wage costs. We have copies of statements made following three wage increases which give detailed estimates of cost increases that should flow from the wage concessions.[1] The statements present separate cost estimates for strip and underground mines. Underground-mine costs being higher than strip costs, wage increases always imposed a greater cost burden on the deep mines. In deference to underground-mine operators (and to secure a price increase above the cost increase for strip mines) the Indiana operators usually sought price increases high enough at least to cover the underground-mine cost increases. Operators in the other districts followed similar practices, basing their decisions on cost considerations with an eye cocked for competitive pressures from other fields that might modify the agreed upon figure.

These pricing practices were never wholly successful. Various impediments prevented their functioning effectively. Several hindrances keep any similar arrangement from working. There is the temptation offered each producer to improve his relative position by "chiselling." Operating under a fixed price in an industry (such as coal mining) suffering from seasonal fluctuations

---

[1] We have statements dated March 6, 1950, August 24, 1955, and the worksheets for a statement issued in the latter half of 1956 (exact date unknown). These statements were part of the Indiana Coal Association's data released to the Indiana University library when the Association disbanded.

and chronic excess capacity, most producers operate at less than capacity. Their inability to adjust price to achieve an optimum profit output rate may frustrate many operators, especially the low-cost producers holding a price umbrella over their high-cost competitors. The slightest weakening of demand triggers price-cutting reactions that can destroy a price arrangement in short order. The larger the group involved, the more difficult it is to hold the members together. Three sellers have only three relationships that need coordinating: A must agree separately with B and C, and B must coordinate his price decisions with C. However, with ten participants the number of such relationships rises to forty-five; with twenty sellers the number increases to 190.

Price shading often takes subtle forms. If the seller is dealing with an unsophisticated buyer he may maintain the f.o.b. mine price at the agreed upon level, but reduce his delivered cost on a heat value basis by "blowing up" his guaranteed B.t.u. above his competitors'. This maneuver, which cannot work with buyers who analyze their coal, depresses a competitor's price as effectively as it would if the price itself were cut — unless the aggrieved seller retaliates by exaggerating *his* coal's heat content. Having already promised the buyer a specified B.t.u. content, however, he is less able to adjust his guarantee than is the outsider trying to capture the account.

Several other price-shading devices are used. Producers who operate their own barge or truck transportation facilities are free to quote delivered prices, disguising in their quotations f.o.b. mine price reductions that are absorbed in the transportation charges. Since operators need not publish their transportation charges, their competitors are unaware that they have broken the price line. In addition, producers operating sales agencies that sell other companies' coal (not members of the association) as well as their own can secure contracts on the outsiders' coal at less than the agreed prices and later transfer the tonnage to one of the violator's own mines.

This practice points up two other forces undermining attempts to stabilize prices. First is the danger that the prices may be set at a high enough level to induce entry by those who were deterred by the previously lower prices. A greater danger is the threat from outsiders who refuse to participate in the pricing arrangement. As long as demand for coal is weak these nonparticipants can effectively nibble away at the association members' accounts, capturing tonnage whenever the group holds the price line. As the number of outsiders increases, the task of preserving the price arrangement becomes more formidable.

Even if the number of nonparticipants is at a minimum, the arrangements face an enforcement problem from the lack of coordination among the different districts. Indiana operators, for example, may effectively regulate interseller price relations in their home state, but in the absence of coordination with other groups they may come into open conflict with Illinois and western Kentucky shippers in Wisconsin or Iowa or Minnesota.

Recognition of their mutual interdependence, together with their rapidly diminishing numbers, led the midwestern operators to lay aside their differences and form a master association early in 1959. The Midwest Coal Producers Institute, Inc., gathers into one organization the discordant factions represented by ten separate trade, traffic, statistical, and labor associations. The previous activities of the associations continue under the direction of committees made up of affected operators. Thus, for example, western Kentucky operators have a labor committee to advise the association's labor commissioner on labor matters directly affecting western Kentucky mines. The traffic activities of each district are regulated in a similar fashion.

The association represents all of the major midwestern producers and most of the medium-sized ones as well. In 1960 it accounted for 87.2 percent of the midwestern production.[2] Only the small mines and a few larger operations not covered by the U.M.W.A.'s labor contract remain outside the association. Little of the nonparticipants' tonnage competes with association members' shipments outside of the T.V.A. market. The absence of nonmember competition and the fewness of sellers within the association provide unique opportunities for coordinating pricing

[2] Based on annual tonnage report issued by Midwest Coal Producers Institute, Inc., compared with the midwest's total tonnage reported in the Bureau of Mines *Minerals Yearbook*.

policies, though there is no evidence that formal coordination presently exists.

To understand fully the individual firm price-making process as well as interseller price relations, we need to divide the analysis into pricing for the long-term contracts and for all other sales. On spot sales and on contract sales where the duration of contract is one year (the standard practice with industrial contracts), the interaction of demand conditions, excess capacity levels, and varying degrees of interdependent pricing have determined price levels and practices more than they have in long-term contract pricing. The basic structure of oligopoly has led to different pricing conduct patterns in the midwest at different periods of time. But through them all (in recent years at least) price-output decisions of each have influenced the others. At times reactions have been suppressed by high levels of demand permitting each firm to supply "its" segment of the market free from attack from competitors. This certainly characterized conditions during the 1947 and 1948 coal boom. At other times oligopoly has led to price wars (1955).

Today in the midwest an uneasy peace exists in the short-term market. Attempts of individual operators to expand tonnage by cutting price and invading competitors' domains continue, but mutual recognition of the disastrous path this course can lead to reduces its incidence under the level existing in the price-cutting 1950's.

At no time has an individual firm assumed effective price leadership. Public price increase announcements by Consolidation Coal Company or Appalachian Coals, Inc., an eastern marketing agency, provided support to demands by midwestern operators to raise prices during the periods of rapid price increases previously referred to. However, no midwestern firm has been accepted tacitly or formally as a price leader. Even today with control centering in fewer hands no price leader emerges. This condition probably stems partly from the lingering suspicions developed in the more turbulent years and partly from the trend toward long-term contracts, a situation which makes price leadership less practicable.

Pricing of long-term contracts, by the very nature of the contracts themselves, is individualized, less subject to interseller coordination. Ironically, this fact may lead to prices capable of guaranteeing the operator higher returns than those stemming from open-market sales, jointly price-determined. If a firm dedicates a mine to supplying a utility for all or most of the mine's life it must charge a price high enough to cover all costs including a normal profit. Discussions with operators indicate that they apparently follow this pricing practice in general. However, there is no consensus on what constitutes a "normal" profit. Whenever possible, operators selling coal under long-term contracts try to de-emphasize mine price in the discussions, shifting attention instead to locational advantages (e.g., water availability for mine-mouth plants) or to reduced freight rates which the operator can negotiate for the coal movement.

Contract Duration and Pricing

We have enough evidence that the pricing of long-term contracts differs from open-market and short-term contract pricing to warrant studying the relationship of contract duration to pricing in more detail. The preceding section indicated the need for full-cost pricing on very long-term contracts to assure the operator a sufficiently high return to warrant investment in the mine furnishing coal under the contract. In Chapter VI we found that prices for coal sold under long-term contracts are more stable than spot sales prices. The same holds true for sales made under short-term contracts with a year's duration. Removing sales from the market for a year lets the firm ride out the summer seasonal slump without damage to its price level.

Industry executives agree with this analysis. Our spot check of a company's sales records confirms it. We studied confidential month-by-month sales realization data for a large Indiana operator for the year 1953.[3] During the seasonal peak in January, his

---

[3] This operator, along with most other Indiana shippers, submitted invoices to the Indiana Coal Association which transferred the data to individual company and consolidated mine sales reports. The Indiana Coal Association submitted copies of these reports to the Indiana University library for research use. We have had access to these reports but are bound by the stricture not to reveal the identity of individual shippers.

average realization per ton was $3.71; in May, the seasonal low, average price had fallen to $3.65. Significantly, none of the price reduction apparently resulted from lower prices to individual customers. Rather, it stemmed from a seasonal change in emphasis as sales to retailers melted with the snow. In both months, the average screenings (coal size used by utilities and industrial users) price was identical, despite a reduction in the amount demanded for these sizes from 66,700 to 33,000 tons.

It would be misleading, however, to assume that seasonal shifts in demand do not influence pricing despite evidence above to the contrary. Demand changes affect price quotations in two ways. First, spot prices for immediate shipment accurately reflect transient demand conditions. The pressure of reduced summertime demand depresses price quotations for shipment during that season. Thus summer lake-cargo shipments, coveted by shippers trying to even out the yearly production cycle, command relatively low prices. Likewise, short-term sales for winter delivery command higher prices. Second, demand shifts react on prices when yearly (or long-term) contracts expire. By tradition, coal contracts formerly terminated on April first when the operators' wage agreement with the Mine Workers' Union expired. New contract prices reflected demand conditions existing on that date. Since by then the summer slump had begun, buyers wielded relatively strong bargaining power. Recently the trend has moved toward the use of contract expiration dates staggered throughout the year, strengthening the operators' position.

A hasty review of the prices charged by any operator would reveal no apparent relation between individual prices and the term of contract under which each sale was made. Some spot sales prices would lie above long-term contract prices; others would fall below them. The type of customer served, the conditions prevailing when contract prices were set, the ability of producers to practice price discrimination would all influence the confused pattern of prices in relation to the term of contract. The only way to determine the effect of contract length alone is to isolate this factor from all others. Table 35 accomplishes this purpose. It abstracts from all variables other than the term of contract. The bids covered only the chemically similar western Kentucky

TABLE 35.   Relation of T.V.A. bid prices and term of contract

Term	Tons/week bid	Tons/year	Average f.o.b. price[a]
12 months or less	41,445	2,155,140	$2.51
13–36 months	19,745	1,026,740	2.64
36 months or more	160,162	8,328,424	2.94

[a] Weighted average price.

N.B. Figures in this table are based on bids submitted on September, 1960, February 9, 1960, September 17, 1959, July 27, 1959.

Source: T.V.A. bid sheets.

no. 9 and no. 11 coals; in almost all cases common freight rates applied to all proposed shipments, eliminating the need for absorbing freight differentials in the quoted f.o.b. mine prices; bids were made at the same time, removing the influence of demand and supply shifts from the price levels.[4]

All three price averages in Table 35 are relatively low, reflecting the stiff competition existing in this market. But as the term of contract increases, the price level likewise increases. Producers desperately needing additional sales are content in the short run to quote a price above their variable costs but not necessarily high enough to cover total costs. As we have seen, when they commit themselves to longer contracts they must adjust to the long-run situation which requires that they cover fixed as well as variable costs.

A word of caution: the relationship between long- and short-term prices previously described represents a normal condition in coal, but by no means does it describe the only relationship which can exist. The normal condition reflects the pressure of excess capacity and unsteady demand which continually operate to damp down prices, especially on spot sales. Soft market condi-

[4] The figures could be slightly distorted if bids for the three contract-length periods were not distributed proportionally over each of the four bid openings. Thus, for example, if a disproportionately large concentration of long-term bids was submitted for a bid invitation when the price level was temporarily inflated, this accident of timing would impart an upward bias to the long-term price level shown in the Table.

tions are an invitation to the sale of spot market coal at depressed prices. Under these conditions (prevailing in the depressed thirties and during much of the 1950's) returns from long-term sales tend to exceed those from short-term sales. But when demand strengthens, the reverse may hold true. Periods of rising prices may find long-term contract prices below those charged on spot sales, but only because of the time lag involved. However, comparing long- and short-term prices entered into *simultaneously,* during periods of generally rising prices, would probably reveal no differential.

Effect of Market Structure on Conduct

Oligopolistic market structures, typified by those existing in the midwestern coal industry, are characterized by interdependence among sellers. We noticed some of the effects of this interdependence in the preceding section. Producers are large enough to make their price-output decisions felt by their competitors. This condition being true, an industry's firms may react in one of three ways. (1) They may agree among themselves on price by means of a formal or informal understanding. (2) Realizing the effect the decisions of each may have on price and supply, they may achieve the same end through "conscious parallelism." Here parallel behavior results "not by agreement . . . but in automatic response to identical stimuli."[5] (3) Failing to establish agreement tacitly or overtly, open warfare may erupt, driving down prices to uncomfortably low levels. Each of the three solutions is equally possible, but the third alternative produces results diametrically opposed to those of the first two.

Having established the oligopolistic character of the midwestern industry in Chapter V, we need now to determine the effect on price of its market structure. Tight market control should reflect itself in more stable prices; price warfare should depress them. Table 36 throws light on the issue. The percentage change in prices is shown for the entire industry, the midwest, and the Fulton–Peoria field for the period 1950–1960. The latter area is a subdistrict within the midwest where production is highly concen-

TABLE 36.   Average realization on coal shipments from midwest, Fulton–Peoria field, and United States, 1950–1960

Year	Midwest average price	Percent change	Fulton–Peoria average price[a]	Percent change	U.S. average price	Percent change
1950	$3.94	−0.3	$3.73[b]	+0.3	$4.84	−0.9
1951	4.00	+1.5	3.79	+1.6	4.92	+1.6
1952	3.98	−0.5	3.69	−2.6	4.90	−0.4
1953	3.79	−4.7	3.68	−0.3	4.92	+0.4
1954	3.62	−4.5	3.60	−2.2	4.52	−8.1
1955	3.49	−3.6	3.67	+1.9	4.50	−0.4
1956	3.67	+5.1	3.89	+6.0	4.82	+7.1
1957	3.84	+4.6	4.12	+5.9	5.08	+5.4
1958	3.83	−0.3	4.15	+0.7	4.86	−4.3
1959	3.88	+1.3	N.A.	N.A.	4.77	−1.9
1960	3.82	−1.5	N.A.	N.A.	4.69	−1.7

[a] Includes only shipments to industrial and utility consumers.
[b] Estimated.
N.A. = Not available.
Sources: Bureau of Mines, *Minerals Yearbook*, 1950–1960; Middle States Fuels, Inc., confidential realization reports.

trated among a few large firms. In 1958 six firms accounted for 96.3 percent of the rail shipments from this field, three large companies controlling 79 percent of the total field output. The field being isolated in northwestern Illinois, removed from the rest of the industry, firms operating there maintain effective market control over a region comprising northern Illinois and eastern Iowa.[6] Its market domination and the structure of mine ownership in the field should reflect themselves in relatively more stable prices than those found in the balance of the midwest. Midwestern prices, in turn, should be firmer when demand falls off than prices in the more atomistic national industry.

Fulton–Peoria prices withstood the pressure from depressed conditions better than both the midwestern and national industry in the post-Korean War period. From the 1951 peak to the trough in 1955 midwestern prices fell 12.7 percent and national prices 8.5 percent, while Fulton–Peoria prices declined only 3.2

[6] The Fulton–Peoria field covers production in Fulton, Knox, Henry, and Schuyler counties in Illinois. Production from rail mines in 1958 amounted to 7,888,032 tons. An additional 45,729 tons came from small truck mines.

percent. Actually Fulton–Peoria prices fell to their lowest point in 1954, but even then its price level was only 5 percent below the 1951 peak. Not only did this subdistrict's prices fall less than those of other areas, they also rebounded faster as well when the cycle turned upward. Prices in the last half of 1955 rose sufficiently to pull the average for Fulton–Peoria mines above the 1954 low, while midwestern and national price averages continued to fall.

How can we account for the relatively concentrated midwest lagging behind the rest of the industry during this downward phase of the cycle? If the comparatively few firms controlling most of the midwestern output had stabilized prices through tacit or explicit agreements, they should have approached in stability the performance of the Fulton–Peoria field. That they not only failed to do so, but lagged behind the rest of the industry as well, indicates that the third behavioral pattern mentioned above may have applied during this period. This in fact was the case. The period 1953–1955 witnessed a bitter price war in many of the midwestern markets resulting from an expansionary drive by western Kentucky operators, and particularly by the Nashville Coal Company. Fortunately for Fulton–Peoria operators their insulated market position protected them from the effects of the war's skirmishes except in a few upper Mississippi River markets which they shared with western Kentucky operators. As a result they emerged from the price war virtually unscathed. Improved coal demand and the purchase of Nashville Coal Company by the West Kentucky Coal Company in 1955 were the proximate causes for the cessation of the war in the fall of that year.

National coal price levels fell again in response to the 1958 business recession, dropping 4.3 percent below the previous year's level. The tightly knit Fulton–Peoria group, however, unresponsive to adverse demand conditions, actually *increased* prices 0.7 percent. By this time merger activity in the midwest had intensified, increasing the concentration ratio to a point approaching the high level existing today. The era of price cutting behind them, midwestern operators reduced prices a scant 0.3 percent in this brief recession. In the final two years under review midwestern prices remained comparatively stable, the 1960 average realization falling only one cent per ton behind the 1958 level, while the national average price level dropped another 3.5 percent. How much of the difference in performance can be attributed to differences in seller concentration and the structure of coal markets on the one hand and to variations in coal demand (midwestern vs. the national industry) on the other is an unresolved question.[7] Quite possibly both forces were at work. If this was true the influence of structural factors on conduct would have been diminished but not removed.

Table 37 relates midwestern prices to the midwestern produc-

TABLE 37.   Relation of midwestern average realization to output rates, 1951–1959

Year	Midwest average price	Percent change	Midwest tonnage (million tons)	Percent change
1951	$4.00	+1.5	95.6	−4.7
1952	3.98	−0.5	83.3	−8.7
1953	3.79	−4.7	83.1	−0.2
1954	3.62	−4.5	77.8	−6.3
1955	3.49	−3.6	88.3	+13.5
1956	3.67	+5.1	94.8	+7.4
1957	3.84	+4.6	91.8	−3.2
1958	3.83	−0.3	86.9	−5.3
1959	3.88	+1.3	89.8	+3.3

Source: Bureau of Mines, Minerals Yearbook, 1951–1959.

tion rate, strengthening the preceding discussion. In the absence of impediments to competition we should expect to find a consistent, fairly close relation between price-level changes and changes in output rates. The table does not reveal this kind of pattern. In the first two years and in 1957 and 1958 when production fell, price reductions lagged. They lagged as well in 1956 and 1959 when tonnage rose. Only during the period 1953–1955

[7] Between 1958 and 1960, e.g., demand for coal increased but on the national level production rose only 1.1 percent while midwestern production increased 5.8 percent.

were price movements responsive to tonnage changes. In the last of these three years prices actually fell while production increased.

How can we account for these price-output patterns? The 1953–1955 relation is easily explained. Prices responded violently to the price war mentioned previously. It was severe enough in early 1955 to prevent the price level from rising despite an increase in over-all demand. The factors influencing the other years are harder to isolate, but two forces doubtless were at work. First was the effect of long- and short-term contracts reducing the short-run elasticity of demand. This factor accounted for price lagging behind output changes in both directions. Also contributing to this lagged relationship — this time, though, only when demand decreased — was the high concentration ratio in the midwest. Attempts to measure the degree to which these two forces operated in the midwestern market would be fruitless. We can do no more than mention them.

Price Discrimination

If the coal industry typifies a perfectly competitive industry, producers would be unable to practice price discrimination. The existence of many firms producing a homogeneous commodity would insure a single price being charged in any market, since if a producer charged a higher price, consumers would quickly shift their allegiance away from the high-price product. That we fail to find a single price prevailing in a given market is not presumptive evidence that firms practice price discrimination in that market. Prices may vary solely because of slight variations in quality that justify attaching different values to different coals. If this is true the coals are not homogeneous, and we may consider them to be separate products.

However, if we observe a seller charging different prices in the same market for the same product we may suspect that he practices price discrimination. Some price variations may result from cost differences, putting them within the pale from the viewpoint of the Federal Trade Commission as well as of the economic theorist. Aside from cases of this type we find abundant evidence of the kind of price discrimination usually associated with the monopolist, i.e., an individual firm selling the same product to different buyers at different price-cost ratios.

How can we account for this phenomenon in the midwestern industry?[8] Several forces are responsible, but by far the most important is the unequal relative bargaining power of the participants in different market situations. In fact the influence of bargaining in price setting is so pervasive that we can construct a "bargaining theory of price" to explain it. Not unlike its counterpart in wage determination, a bargaining theory of price looks to the relative bargaining strength of sellers and buyers as determinants of the price "bargain." But this strength is not unlimited. We should view bargaining power as operating within bounds set by the underlying market supply and demand conditions. Price is free to move between some kind of upper and lower limits in response to relative bargaining positions.

The phenomenon is not restricted to the coal industry by any means. Most industrial goods carry list prices which may or may not coincide with selling prices. Even in the steel industry, relatively invulnerable to downward pressure on list prices, periods of slack demand bring with them "price-shading" practices that stem from the shift in bargaining power away from sellers. The effect of buyers' and sellers' relative bargaining strength is particularly evident in the selling of auto and truck tires. Most industrial tire salesmen have a seemingly endless string of discounts tucked away, ready to flash before hard-bargaining buyers.[9]

What forces influence the relative strength of participants in coal markets? Most important on the buying side is the buyers' bargaining power which is conditioned by his size. Large electric utility buyers possess sufficient strength to buy coal under the most favorable terms. Small buyers — especially those requiring coal during the peak winter months — are apt to be in a relatively weak bargaining position. The bargaining skill of the negotiators of a price bargain is an equally important determinant of bargaining power. A well-informed, sophisticated, hard-driving

[8] The practice of price discrimination in the coal industry is not confined solely to the midwestern sector. The ensuing discussion would apply equally well to the balance of the industry as well.

[9] The bargain theory of prices perhaps reaches its ultimate in the colorful contest between seller and buyer in small, Latin American outdoor markets.

coal buyer purchasing a small supply of coal may possess enough skill to overcome the weakness of his low consumption rate. In addition, otherwise weakly situated sellers may strengthen their position by selling to users who reciprocally supply their products to the mine.

All of these factors show up in an analysis of Table 38 which

TABLE 38. Screenings prices on shipments from Indiana mine to Indianapolis customers, August, 1953

Customer	Tonnage	Price
John Wachtel Company	207	$3.40
Indianapolis Power and Light Company	4,638	3.41
City of Indianapolis Schools	51	3.41
R.C.A. Manufacturing Company	125	3.50
Eli Lilly Company	1,251	3.60
U.S. Rubber Company	352	3.75
Stewart Warner Company	54	3.80
Mutual Milk Company	51	3.90

Source: Materials furnished by courtesy Indiana Coal Association through Lilly Library, Indiana University.

lists prices charged by an Indiana operator to industrial and commercial users located in Indianapolis. The range between the high and low prices charged was fairly substantial. Indianapolis Power and Light Company's great buying power explains its low price. Its buying power results from its sophistication in purchasing and from the size of its purchases which offered suppliers the opportunity to secure overhead-spreading "backlog" tonnages. Harder to account for is the even lower price of John Wachtel Company, a small industrial user, although the relative skill of the opposing salesman and purchasing agent could have been influential. The City of Indianapolis coal sold at a low price because it was purchased under a sealed bid arrangement, which in a soft market tends to depress prices. The company's salesmen were unable to exert personal influence to elevate the price, as they did to Stewart Warner and Mutual Milk. The fact that bid prices become public knowledge undoubtedly accounted for the identity of the City of Indianapolis and Indianapolis Power and Light Company prices. Any lower, publicly posted price might have forced the utility to demand a similar concession. The operator's bargaining position in sales to United States Rubber Company was strengthened by the power of reciprocity, his mine being a large consumer of the rubber company's products. This reciprocal buying power accounts for the higher price charged to United States Rubber, for example, than to R.C.A.

Review of the defunct Coal Trade Association of Indiana's price records indicates that the discriminatory pricing revealed in Table 38 represented a normal pattern. Individual producers charged numerous prices for the same product in the same market to different buyers. Some of the prices, of course, were not discriminatory since the contracts may have contained different conditions, e.g., different lengths, amounts, etc. Still, many of the sales covered shipments of similar tonnages to like consumers at different prices. Many of these buyers were industrial consumers who usually purchased coal under one-year contracts. The degree of disparity among prices to different consumers appears to be partly a function of the intensity of competition within the various submarkets. Less discriminatory pricing existed in the early post-World War II seller's market, for example, than there was in the more intensely competitive period covered by Table 38.

Influence of Coal Union on Pricing

A view prevails in some quarters that the United Mine Workers' Union, by controlling output, affects coal prices. Before leaving the area of pricing conduct, we need to lay this misconception to rest.

Adelman offers a typical exposition of this view.[10] He argues that "since the end of World War II, at least, Mr. [John L.] Lewis — as the grand coordinator of the coal industry — has done for it what it cannot do for itself because of the antitrust laws or even without them. By controlling the input of labor he has controlled the output of bituminous coal (other than strip-mined)

[10] M. A. Adelman, "Steel, Administered Prices and Inflation," *Quarterly Journal of Economics*, LXXV (Feb. 1961), 16–40.

in such a way as to maximize the revenues of the industry." [11]
He arrives at this conclusion apparently after a careless reading
of C. L. Christenson's article, "The Theory of the Offset Factor,"
dealing with the coal industry.[12] Christenson quoted the *Mine
Workers' Journal*'s declared intention to "stabilize" the industry
by regulating coal production. The *Journal* declared: "A divide-
the-orders, share-the-work program is now a 'must' in the soft
coal industry. Since the soft coal industry on its own accord cannot get together to exhibit the business acumen necessary to
protect its employees and the business and population of the
mining communities, the duty of performing this public service
devolves upon the only stabilizing force the industry has ever
known, the U.M.W.A." [13]

Misreading Christenson's article, Adelman commits three errors. He evidently assumes that (1) the Mine Workers' three-
day-work-week program in mid-1949 was designed to control supply sufficiently to give coal operators monopoly profits that could
support higher wages; (2) this restrictive program continues today; and (3) the nonexistent control of labor input "was en-
trusted by both sides to Mr. Lewis." [14]

By the summer of 1949 the "bloom" was off the postwar coal
market. Demand was falling; mines were operating far below
capacity; inventories were piling up in customers' storage piles.
Lewis was in an obviously weak bargaining position. The situation demanded action. Lewis responded by imposing on the industry a mandatory three-day work week in July, 1949.

Adelman's analysis implies that the key to this move was the
union's ability to create monopoly profits. Demand being inelastic, the restriction of output would move the industry up its
demand curve. The resulting monopoly profits would provide the
wherewithal to meet the miners' demands for higher wages.

There is a more likely and far less subtle explanation for the

---

[11] *Ibid.*, p. 22.
[12] C. L. Christenson, "The Theory of the Offset Factor," *American Economic
Review*, XLIII (Sept. 1953), 513–547.
[13] United Mine Workers of America, "Welfare and Retirement Fund: Four
Year Summary and Chronology" (Washington, D.C.), 1951, p. 33.
[14] Adelman, "Steel, Administered Prices and Inflation," p. 28.

move. Reducing the work week lowered storage piles, putting a squeeze on consumers who were hard pressed to find fuel to meet their heating and steam-generating needs. This kind of pressure had worked to Lewis' advantage in previous years. There was no reason to believe it would not work again, provided coal stocks were reduced to more normal levels. The work limitation scheme effected this reduction, strengthening Lewis' hand when, during that winter, he called a full-scale strike. Additionally, it assuaged the miners by sharing the work, a hoary union device unrelated to attempts to maximize employers' revenues.

Whatever the motives for the control of labor input may have been, the episode is of historical interest only, since the input controls have long since been abandoned. The mandatory three-day work week began in mid-July, 1949, lasting through the balance of July, August, and up to September 19 when a full-scale strike began. The balance of that fall and winter was an unstable period with brief periods of work interrupted by the "half-strike" technique of restricted work weeks and by full-fledged strikes. The operators and the union finally reached a general agreement on March 5, 1950, ending one of the bitterest periods of labor-management relations in an industry long divided by labor strife. The cold-shower effect of this divisive period led to a fresh new approach to labor relations in the industry. The result has been an unprecedented period of peaceful labor relations. More important for the present discussion is the fact that the control of labor input embodied in the three-day work week ended with the stormy close of 1949–1950 contract negotiations, and has never been reinstated.

Adjustment of Product

In a solid state before mining, coal fractures into different-sized particles when it is mined and prepared. The tipple operation consists partly of sorting out the different sizes produced. Some of the sizes are desirable, others, undesirable. The problem is similar to the situation facing petroleum refiners who must balance the output of a myriad of refined products with the shifting demand for those products. The "products" coal producers are concerned with are different sizes of the same mineral, not, as in petroleum refining, separate commodities with vastly different viscosities.

Consumer classes differ in their demand for various sizes. Their preferences result partly from custom, but primarily from the fuel demands of their burning equipment. Demand patterns for the different sizes shift temporally, both in the long run and in the very short run, creating significant pricing problems for the producer.

Fortunately for the producer, the size-balancing problem today is not as severe as it formerly was. Before demand for screenings by utility and industrial users increased, coal demand centered largely on the large lump and egg sizes.[15] These sizes constituted from 40 to 50 percent of the average producer's output. The demand for screenings being limited and their supply governed by the jointly produced large sizes, the fine sizes often found their way into distress markets. In recent years the trend has moved away from the demand for large sizes as the locomotive fuel and retail markets have declined. The most serious remaining joint-product problem is the one presented by the industrial or retail customer demanding stoker coal. This is a superior product from which the very fine coal (called carbon) has been removed. This product commands a higher-than-average price. But producing stoker coal automatically produces carbon. Since this fine coal is difficult to burn, the operator faces a limited market for the product. Thus he gambles that a high stoker price will counterbalance an uncertain carbon price.

With this joint-product problem as well as with the problem mentioned earlier of balancing the output of screenings and large coal, producers face a continual handicap: the reluctance of consumers to shift from relatively high-priced sizes to sizes whose prices are temporarily depressed. Denied the cushioning effect of the substitutability of coal size, prices for some sizes often have fallen far below the market average. In the long run, however,

---

[15] Fine coal falls into two categories: screenings and carbon. Screenings refers to coal small enough to pass through a screen with perforations approximately three quarters of an inch to an inch and one half in diameter. Carbon is a "resultant" product small enough to pass through tiny holes from one-fourth inch to 28 mesh (28 holes per square inch) in diameter. Lump and egg sizes pass *over* screens with perforations 2 to 4 inches in diameter.

there is more substitutability among sizes as consumers install new burning equipment capable of burning lower-priced sizes. Thus in the 1920's and 1930's, industrial and commercial users switched from hand-fired furnaces requiring relatively high-cost egg coal to automatic burning equipment using screenings. Recently many companies have installed furnaces capable of burning carbon to take advantage of its chronically low price. These long-run developments have stimulated demand sufficiently to prop up these sizes' sagging prices.

The pressure from the joint-product problem has been relieved from another direction as well. The demand for coal having switched from the large sizes to screenings, producers can balance sizes by crushing large coal into finer coal to meet the demand. Formerly, if demand for lump and egg coal exceeded the available supply, operators were powerless to correct the imbalance without increasing the already excess supply of screenings. Today they may be limited only by the lack of sufficient crushing capacity which the addition of extra crushers easily rectifies.

Many of the new mines, captive to electric utilities, produce only one size, crushing all coal into a product with a maximum size specified by the utility. This practice eliminates the joint-product problem. It also materially reduces operating and capital costs. A preparation plant geared to produce a wide variety of sizes requires an elaborate system of sizing screens to separate the total input into the requisite "products." Loading the many sizes into railroad cars requires a costly system of supporting railroad track facilities. Loading a single size removes the need for much of this investment; the tipple is simplified, and track layout is reduced to one or two tracks from the usual five or six.

CHAPTER VIII

Market Performance in the Midwestern Coal Industry

This chapter studies market performance in the midwestern industry. By performance we mean the "strategic end results of market adjustments engaged in by sellers" serving as "the crucial indicator and measure of how well the market activity of enterprises contributes to the enhancement of general material welfare."[1] We seek to link performance to structure and conduct when they are causally connected.

The study of "end results" falls into several parts. First, we analyze profits rates in the midwest, comparing them wherever possible with earnings in other sections of the industry. Later, we study the degree of productive efficiency existing in the midwest, the pattern of income distribution, and the standards of social and economic performance achieved by the midwest in natural resource conservation. As data underlying a study of these performance dimensions we investigate the effect on costs of existing excess plant capacity, the impact of accomplished mergers on efficiency, and the structure of coal royalty payments.

As with price information, earnings' data in the coal industry are hard to obtain. Most of the companies in the industry being small, few list their stock on major stock exchanges. As a result only a handful publishes annual earnings' statements. Fortunately for an investigator of the industry, midwestern companies with listed securities account for proportionally more of their region's production than is true of companies operating in other sections of the industry. Unfortunately, however, since these companies

[1] Bain, *Industrial Organization,* p. 340.

are larger than most, the results are probably not representative of industry earnings. The question of how to treat the percentage depletion cost deduction presents an additional handicap in analyzing industry profits. The Internal Revenue Service treats percentage depletion as a cost item, deducting it from earnings when reporting industry profits in its *Statistics of Income*. Coal operators, correctly viewing the deduction not as an item of cost but as a tax-saving device, add percentage depletion to reported profits in their financial statements. The diverse treatment of this deduction makes a comparative analysis of I.R.S. and corporately published earnings' statements a tricky business.

Prewar Earnings Comparison

Cost and realization data submitted to the government during periods of federal intervention in the industry's affairs provide a basis for comparing earnings in various industry sectors. Ideally these figures should compare earnings as a percentage of net worth if they are accurately to reflect investment values. Unfortunately the only early data available report earnings as pre-tax operating margins expressed in cents per ton. They provide only a rough guide to comparative rates of return, and need to be read with their limitations in mind.

Table 39 compares operating margins for midwestern mines with those of most of the rest of the industry for the period 1936–1941. Districts 1 to 8 include all of the major coal districts in the Appalachian region and together with the midwest account for approximately 90 percent of United States' production. Costs are determined uniformly and include all but a few production and nonproduction cost items.[2] Excluded costs amount to approximately 5 cents per ton and their exclusion, applicable to all mines, does not affect the profit margin comparisons.

Neither the midwest nor the rest of the industry grew fat on profits during the six-year period under review although the midwest consistently outpaced Districts 1 to 8. Weighted average

---

[2] Excluded from reported costs were: taxes on unallocated coal lands, or allocated coal lands which would remain unmined for thirty years, interest, bad debts, demurrage charges, costs incurred by idle and abandoned mines and mines under development, and the costs of maintaining company houses.

TABLE 39. Costs, realization, and margins for bituminous coal mines, Districts 1 to 2, 1936–1941 (dollars per ton, f.o.b. mine)

Districts	Year					
	1936	1937	1938	1939	1940	1941
1 to 8						
Average realization	$1.81	$2.00	$1.96	$1.91	$1.92	$2.25
Average cost	1.89	2.10	2.13	2.00	1.94	2.20
Operating margin	−0.08	−0.10	−0.17	−0.09	−0.02	0.05
9 (Western Kentucky)						
Average realization	$1.28	$1.37	$1.33	$1.35	$1.39	$1.57
Average cost	1.38	1.57	1.48	1.41	1.42	1.51
Operating margin	−0.10	−0.20	−0.15	−0.06	−0.03	0.06
10 (Illinois)						
Average realization	$1.69	$1.76	$1.73	$1.65	$1.69	$1.81
Average cost	1.63	1.74	1.72	1.62	1.59	1.68
Operating margin	0.06	0.02	0.01	0.03	0.10	0.13
11 (Indiana)						
Average realization	$1.57	$1.62	$1.65	$1.59	$1.62	$1.75
Average cost	1.46	1.63	1.66	1.55	1.53	1.61
Operating margin	0.11	−0.01	−0.01	0.04	0.09	0.14
9, 10, 11 (weighted avg.)						
Average realization	$1.62	$1.68	$1.65	$1.59	$1.62	$1.75
Average cost	1.56	1.69	1.66	1.55	1.53	1.61
Operating margin	0.06	−0.01	−0.01	0.04	0.09	0.14
Midwest margin in excess of margins for Dist. 1–8	$0.14	$0.09	$0.16	$0.13	$0.11	$0.09

Source: Extension of Bituminous Coal Act of 1937, Hearings on H.R. 356, H.R. 1454, and H.R. 2296, U.S. House Committee on Ways and Means, 78th Congress, 1st Session, June 21–July 5, 1943, pp. 21, 22.

margins for the midwestern districts ran from 9 to 14 cents per ton higher than margins for nonmidwestern districts; however, the three districts did not contribute equally to the midwest's improved position. Western Kentucky's dismal performance lagged far behind the rest of the midwest. Profits were lower not because of high costs but as a result of depressed prices. Indeed, with much of the state operating nonunion mines, its costs were among the lowest in the industry. Obviously it was the cost-price spread that determined profits and not the level of absolute costs alone.

We may find in western Kentucky's laggard performance an

explanation for the different earnings' rates existing in the industry. The high degree of seller concentration now existing in the midwest did not prevail in the late 1930's. But the relative isolation of the midwest from the rest of the industry was as important then as it is today. And Illinois and Indiana operators who for the most part washed and prepared their coal had their coal reasonably well accepted in a fairly extensive protected market area. Moreover, by washing their coal many operators differentiated their product sufficiently to reduce slightly the existing bitter price competition. On the other hand, few western Kentucky operators washed their coal during this period. This had the dual effect of limiting market acceptance of their product and preventing operators from shifting the competitive emphasis from price to "product." The only basis for appeal left open to western Kentucky operators was price as they vainly sought to develop markets for what was essentially a substandard product.

Listed Companies' Profits

We can get a measure of recent earnings for midwestern companies as a percentage of invested capital only by studying profits for companies whose stocks are listed on national or regional stock exchanges. These firms must submit financial reports to comply with exchange regulations, thus providing the information necessary for our calculations.

Table 40 summarizes this information. The midwestern figures account for nearly two thirds of 1960 midwestern production. A small percentage of the production listed as midwestern tonnage comes from the few nonmidwestern mines operated by the seven midwestern firms. Including data for these mines (and the figures cannot be segregated) distorts the picture very little. If anything, including them reduces the midwestern rate of return below what it would be in their absence. Truax–Traer, for example, reported that for the three-year period ending April, 1959, its West Virginia property (later disposed of) earned an average return on investment of only 0.7 percent.

For the ten-year period covered by the table the midwestern companies earned an average rate of return of 8.75 percent, eastern listed companies, 6.93 percent. The midwest's advantage was

TABLE 40. Earnings as percent of net worth: midwestern and eastern listed companies, 1951–1960 (dollar figures in millions)

Coal region	Year									
	1951	1952	1953	1954	1955[c]	1956	1957	1958	1959	1960
Midwestern Listed Cos.[a]										
Net worth	$125.0	$130.6	$134.7	$127.5	$149.7	$187.2	$208.6	$224.4	$240.1	$256.2
Net income	16.8	13.2	9.4	5.6	10.2	18.3	19.5	18.4	21.3	24.5
Profit as percent of net worth	13.42	10.10	7.01	4.39	6.84	9.78	9.36	8.20	8.83	9.57
Tonnage (millions)[d]	40.6	36.5	34.5	31.2	40.2	54.7	55.7	51.6	59.0	61.3
Profit per ton[e]	0.41	0.36	0.27	0.18	0.25	0.34	0.35	0.34	0.36	0.39
Eastern Listed Cos.[b]										
Net worth	$296.9	$300.9	$320.6	$325.6	$288.1	$410.4	$444.7	$442.0	$450.3	$443.3
Net income	29.1	22.0	21.4	14.9	21.1	35.6	41.6	24.7	23.6	21.2
Profit as percent of net worth	9.79	7.31	6.67	4.59	7.28	8.67	9.36	5.60	5.24	4.78
Tonnage (millions)	54.1	47.3	45.2	39.6	49.5	66.2	63.8	48.6	55.8	52.3
Profit per ton	0.53	0.46	0.47	0.37	0.43	0.54	0.65	0.51	0.43	0.41

[a] Includes following coal companies: Peabody, Truax–Traer, United Electric, Ayrshire, West Kentucky, Old Ben, and Ziegler.

[b] Includes following companies: Consolidation, Island Creek, Rochester and Pittsburgh, Westmoreland, Stonega and, from 1956–1960, North American. Pocohontas Coal Co. is included for the period 1950–1954; thereafter its figures are grouped with Consolidation Coal Co., into which it was merged.

[c] Includes only 8 months of Peabody's earnings and ⅔ of its year-end net worth.

[d] Figures include several million tons of nonmidwestern production.

[e] Profits are figured on an after-tax basis and include excess percentage depletion.

Source: Moody's Industrial Manual, 1950–1960.

less pronounced during the first half of the period when intense rivalry reduced prices and correspondingly depressed profits. From 1951 to 1955, midwestern earnings averaged 8.35 percent of net worth while eastern mines earned 7.13 percent. The 1956–1960 period saw the midwestern rate of return rise to 9.15 percent, the eastern figure declining to 6.73 percent.

Though the average midwestern rates of return were generally higher than the east's, eastern mines earned larger profits per ton. The discrepancy in these two measures of profits probably stemmed either from the existence of more excess capacity in the east or larger capital requirements per ton of output, or a combination of the two. Note, however, that though midwestern profits per ton were lower than eastern company earnings, they were much more stable, especially in recent years. In the years following 1955, when the midwestern price war ended and the movement toward greater seller concentration intensified, profits averaged between 34 and 39 cents per ton. This profit stability follows from the price stability shown in Chapter VII.

Profits for both midwestern and eastern companies with listed securities surpassed those earned in the balance of the industry. Despite the previously noted difficulties reconciling *Statistics of Income* data with statistics for individual firms from Moody's *Industrial Manual,* we calculated average industry profits for the period 1951–1958 to be 4.8 percent of net worth.[3] We corrected

[3] Internal Revenue Service, *Statistics of Income, Corporations, 1951–58.* Profits were computed on an after-tax basis. They were adjusted to include excess percentage depletion on the assumption that this charge represents not a true cost but a tax relief measure. Since the *Statistics of Income* does not segregate cost and percentage depletion, we devised a means of estimating these amounts. We determined the average cost depletion deduction for mines not earning net incomes by subtracting from the total industry depletion expense, deductions for mines earning net incomes (reported separately in *Statistics of Income* data). We next estimated the cost per ton of sustained depletion for mines suffering losses by dividing their gross revenue by the average industry realization, assuming their realization would not deviate from the national average. This sustained depletion rate per ton was applied to the tons produced by mines earning net income, and the resulting dollar amount was subtracted from the profitable companies' total depletion charge. The balance was assumed to include only excess percentage depletion. This excess was added to after-tax earnings to provide a profit figure calculated in the same way the listed companies reported their earnings. We are justified in assuming that the loss

the reported figures to include excess percentage depletion to make them comparable with the listed companies' data. During the same eight-year period the eastern companies in Table 40 earned an average return of 7.39 percent, the midwestern firms, 8.63 percent. Since both the eastern and midwestern listed firms are comparatively larger than other firms in the industry, we may suspect that size is a determinant of profitability.

It is interesting to compare the earnings' performance of the seven listed midwestern companies with other firms in the economy. No attempt was made to conduct an industry-by-industry comparison. We note, however, that the five hundred largest industrial corporations in the annual *Fortune* survey earned an average return on invested capital of 11.2 percent for the period 1956–1960. During the same five-year span, the seven midwestern coal companies earned 9.15 percent. Included among the five hundred companies are many firms operating in highly concentrated industries. It may come as a surprise that coal companies, operating in what has traditionally been viewed as a depressed industry, earned profits at a rate not appreciably lower than the average of the nation's largest industrial firms.[4]

Relation of Company Size to Profits

The preceding section suggested that corporate size may influence earnings. We noted that the earnings' performance of the larger companies, which listed their securities on local and national stock exchanges, outpaced the balance of the industry. This section throws additional light on the relationship of size to profits. Again the available information is incomplete, covering only the performance of western Kentucky operations during a three-year span. But unlike the previous data which were restricted to the larger firms, here we have information on a repre-

---

corporations deducted only cost depletion since percentage depletion is earned only when profits are earned. This calculation is subject only to the possible error of its assumptions, which are that unprofitable and profitable mines incur identical cost depletion deductions and that they both sell coal for the same average price.

[4] In 1960 the 7 midwestern firms earned a return on net worth of 9.57 percent, less than one half of one percent lower than the average of the "500 largest" for that year.

sentative cross section of firms, small and large alike.[5] The figures include from 65 to almost 100 percent of the district's total tonnage, excluding figures for some small mines which fail to file balance sheets with the Kentucky Department of Revenue and several large mines operated by firms with multistate interests. Since most of the latter firms file consolidated balance sheets, net worth attributable to their Kentucky operations cannot be separated from the total. Excluding these two groups, however, probably does not distort the accuracy of the figures.

Table 41 confirms the previous impression that larger firms earn larger profits than their smaller rivals. For each of the three years studied, the pattern is the same. Firms in the lowest-tonnage class earn the lowest profits (negative in each case); medium-sized firms reap positive, but modest profits; companies in the top two tonnage groups earn the highest returns. Though this relationship is clear-cut it is not linear. Earnings for firms in the largest-sized group, for example, are less in each instance than those for companies in the 500,000- to 1,000,000-ton category. But the sample is too small to generalize about this result. It may be due, not to diseconomies of scale, but to the fact that firms in the largest-sized group operate underground mines which are inherently less efficient than the strip mines which account for much of the production in the next smaller group.

The table tells us something else. There appears to be a break around the 500,000-ton mark between mines making reasonable returns on investments and those earning submarginal returns. We can only surmise that better reserves, better sales connections, and greater productivity from improved technology for the larger firms contribute to this dichotomy. Our study of costs in Chapter V partially supports this view.

Finally, Table 41 reveals a sharp reversal in western Kentucky's fortunes in the post-World War II period compared with her

[5] I am indebted to Professor W. W. Haynes of the University of Kentucky and to the Kentucky Department of Revenue for permission to study the confidential income and balance sheet data submitted by Kentucky coal operators to the Department for Kentucky state income tax purposes. A summary of these financial reports for each reporting company forms the basis for our analysis. Permission to use the figures was granted with the understanding that the data would be grouped to avoid disclosure of individual company figures.

TABLE 41.   Relation of earnings to company size, western Kentucky operations, 1952–1954

Annual company production (tons)	No. of cos.	Total tonnage (thous.)	Net profits[a] (thous.)	Net worth (thous.)	Profit as percent of net worth
		1952			
−100,000	16	611	$−154	$ 308	−50.0
100–200,000	3	398	−94	1,300	−7.3
200–500,000	4	1,288	77	2,289	3.3
500–1,000,000	5	3,171	1,386	6,622	20.9
+1,000,000	5	12,820	5,094	33,255	15.3
Total	33	21,288[b]	$6,309	$43,774	14.4
		1953			
−100,000	9	361	$ −32	$ 165	−19.4
100–200,000	3	377	46	1,235	3.7
200–500,000	—[c]	—[c]	—[c]	—[c]	—[c]
500–1,000,000	9	5,940	1,811	16,076	11.3
+1,000,000	2	9,510	2,297	22,505	10.2
Total	23	16,188[d]	$4,122	$39,981	10.3
		1954			
−100,000	7	151	$ −23	$ 91	−25.3
100–200,000	4	499	47	1,214	3.9
200–500,000	2	681	45	1,207	3.8
500–1,000,000	8	6,495	1,469	16,842	8.7
+1,000,000	2	7,638	1,229	20,635	6.0
Total	23	15,464[d]	$2,767	$39,989	6.9

[a] Includes postfederal income tax profits plus excess percentage depletion.

[b] Adds to slightly more than reported tonnage shown in Bureau of Mines' records.

[c] Combined with figures in next largest group to avoid disclosure of individual company data.

[d] Accounts for 76 percent and 65 percent, respectively, of total western Kentucky tonnage for 1953 and 1954.

Source: State of Kentucky Department of Revenue, confidential tax records.

dreary prewar performance. Recall that in five of the six years covered by Table 39, District 9 operations suffered losses. The increasing acceptance of western Kentucky coal as washeries were installed, cost reductions following the introduction of large-scale mining techniques, and the accessibility of western Kentucky mines to cheap water transportation all contributed to the district's postwar profit improvement.

Effect of Excess Capacity

The extent to which existing plant and equipment is used has such a pervasive influence on efficiency, costs, and resource allocation that it requires special study. We skirted the problem of excess capacity in the discussion of entry barriers. The present discussion, however, calls for deeper treatment of the subject.

John P. Miller asserts that excess capacity exists when "there is momentarily a greater willingness to supply product over a certain price range than there would be if time were allowed to make all equilibrium adjustments." [6] That is, there is excess capacity "if the quantity of factors associated with the industry would have been less if the present prospects of future demand and costs had been foreseen." [7] In coal, excess capacity is said to exist when mines operate, on the average, less than 280 days a year. This standard, erected by the Bureau of Mines, purportedly represents the number of days the average mine can operate, allowing for idle time for Sundays, holidays, and mine disabilities.[8]

The standard is deficient on several counts. First, it fails to distinguish between operating conditions in underground and strip mining that result in different output potentials for the two mining methods. Underground operations can be expanded by using existing equipment on second- and third-shift operations, or even on Sundays if conditions demand it. A strip mine lacks this flexibility, its output being limited by the capacity of its stripping equipment. Normal operations call for shovels or draglines to operate around the clock for five to six days per week, uncovering enough coal to load on four- to five-day shifts. During periods of peak demand, stripping equipment can work three shifts on

[6] John P. Miller, "The Pricing of Bituminous Coal," in *Public Policy*, ed. C. J. Friedrich and Edward S. Mason (Cambridge: Harvard Univ. Press, 1940), p. 151.

[7] *Ibid.*, p. 151.

[8] Capacity levels should really be shown, not as a static function of the amounts existing plants are capable of producing by working a given number of days per year, but as schedules of amounts potentially supplied at different prices.

Sundays to supply one additional tipple shift. The present method of determining capacity figures fails to account for the divergent abilities of strip and underground mines to expand short-run supplies. Is it logical to use the 280-day figure to calculate industry capacity without first determining the *daily* capacity potential? Without this added piece of knowledge capacity figures lack meaning.

A second limitation — related to the first — is the invalid assumption that strip mines can operate as many days as underground mines. As indicated, a strip mine depends completely on the full-time availability of one excavating machine. A breakdown of this large and complex machine can completely halt all mining operations, unlike the situation in deep mines where a large number of independently operated machines function. Furthermore, adverse weather conditions during cold and rainy seasons hamper strip operations which, again, unlike deep mines, are conducted in the open. These factors combine to reduce available running time in strip mines and call for reducing capacity limits to 260 tipple-operating days per year. We make this amendment to Bureau of Mines' capacity figures in subsequent capacity calculations.

Even using the amended measuring rod (which reduces capacity totals), there has been persistent overcapacity both in the national and midwestern coal industries. Many factors contribute to chronic excess capacity in any industry. Important are (1) technological improvements in productive processes that stimulate expansion of new facilities before all existing obsolete facilities are retired; (2) monopolistic and oligopolistic pricing practices that maximize profits at below-capacity output levels; (3) the indivisibility of productive units requiring large increments to be added; (4) seasonal demand; (5) frictional and secular changes, such as shifts in demand from one product to another or geographical shifts of markets.[9] The excess-capacity problem is aggravated if this condition combines with the immobility of resources in the affected industry to convert what might be a temporary maladjustment into a chronic condition.

[9] See John A. Shubin, *Managerial and Industrial Economics* (New York: Ronald Press, 1961), pp. 173–174.

The factors mentioned in items 4 and 5 above appear to have been most relevant for the coal industry. The influence of seasonal demand is obvious enough to require no further comment. We have previously described the contraction in coal demand resulting from the substitution of competitive fuels, a condition that has continued unabated for four decades. The immediate effect of this development on producers is to force them to contract output, moving them up their short-run cost curves. Profits decline and in many cases become negative. In the short run, firms hang on stubbornly, if they can cover variable costs, waiting for a brighter day to dawn. Each firm tries the apparently easy remedy of cutting prices to recapture lost tonnage. Each learns in dismay that competitors follow suit, nullifying the price-cutter's temporary advantage.

Failing to correct the problem from the demand side, operators turn their attention to costs. Throughout the 1920's and early 1930's, as we have seen, many producers tried to widen narrowing profit margins (or turn red ink into black) by converting mines from union to nonunion operations, driving down the wage level in the process. Operating on lower short-run cost curves, producers pushed prices even lower in a desperate attempt to expand short-run output. Many of them succeeded, though in the process prices were sufficiently depressed to eliminate profits for all but the lowest-cost producers. And financial stringency generally prevented the expansion of plant scale which might have strengthened each firm's competitive position by moving it down its long-run cost curve. Eventually, unable to recover fixed costs, many submarginal firms finally left the industry, reducing excess capacity.

The response to reduced demand of midwestern operators has varied somewhat from the pattern described above. Faced with a firmly entrenched union, Illinois and Indiana operators have been unable to meet the cost-price squeeze by reducing wages, though in less well-organized western Kentucky, "wage competition" has always existed. In recent years, though, as large unionized operations have expanded production in western Kentucky, the percentage of nonunion tonnage mined there has declined. Nevertheless, many smaller operators (and a few medium-sized ones) from this district continue to stave off the effects of excess capacity by wage cutting.

Midwestern vs. United States' Excess Capacity

There is evidence that the above factors and the entry barriers discussed in Chapter VI combine to reduce excess capacity in the midwestern sector below that prevailing in the rest of the industry. Table 42 reveals that this relationship has not always prevailed. But the trend in favor of the midwest is unmistakable.

The table indicates the extent of excess capacity indirectly by expressing its counterpart — the average operating rate. The latter is defined as the ratio of the number of days mines operated to the number of available work days (capacity). Capacity figures are weighted averages of strip- and underground-mine capacities, determined on the basis of 260 and 280 work days annually for the two extractive methods respectively. Excess capacity then is the difference between 100 percent and the operating rate.[10]

The midwestern mines have shown steady improvement in the reduction of excess capacity. In the first period, 1928–1933, a heavy burden of excess capacity pressed down on the midwest as this sector of the industry produced at an operating rate of only 53.6 percent. This was the era which witnessed an intensive nonunion drive in the east, enabling the low-wage districts in West Virginia and eastern Kentucky to capture new markets at the expense of the unionized, high-wage cost, northern mines. The operating rate for the United States as a whole was 64.5 percent.

The widespread extension of unionism in the industry following John Lewis' organizing drive in 1933–1934, together with increased mechanization in Illinois and Indiana, redressed the previously existing cost imbalance and improved the midwest's competitive position vis-à-vis the rest of the industry. This improved position reflected itself in a rising operating rate, excess capacity being reduced from 46.4 to 38.7 percent. The improve-

---

[10] We measure capacity in terms of a "feasible" output rate, ignoring for the time being the question of whether or not this rate can be achieved at minimum costs.

TABLE 42. Capacity, number of days worked, and operating rate, midwestern and United States' mines, 1928–1960

Year[a]	Indiana			Illinois			Western Kentucky[b]			Midwest			United States		
	Capacity (days)	No. days worked	Oper. rate per cent	Capacity (days)	No. days worked	Oper. rate per cent	Capacity (days)	No. days worked	Oper. rate per cent	Capacity (days)	No. days worked	Oper. rate per cent	Capacity (days)	No. days worked	Oper. rate per cent
1928–33	273	156	57.1	277	146	52.7	280	148	52.9	276	148	53.6	279	180	64.5
1934–41	269	177	65.8	276	168	60.9	279	155	55.6	274	168	61.3	278	188	67.6
1946–55	268	192	71.6	274	189	69.0	271	190	70.1	272	190	69.9	275	198	72.0
1956–60	266	212	79.7	271	211	77.9	269	207	77.0	269	210	78.1	274	196	71.5

[a] The abnormal wartime years are excluded.
[b] 1928 data unavailable.
N.B. Figures for the districts and the United States for each subperiod are unweighted averages, but the midwestern figures are weighted by the average annual tonnage produced by each district in the subperiods.
Source: Bureau of Mines, *Minerals Yearbook*, 1928–1960.

ment carried forward into the postwar period when the midwest's operating rate approached the rate enjoyed by the rest of the industry.

In the period 1956–1960, when the merger movement accelerated, excess capacity continued to be reduced, the midwest achieving an operating rate of 78.1 percent. All three districts shared in the improvement. Meanwhile the operating rate for the rest of the industry declined modestly, falling to 71.5 percent, 6.6 percent below the midwest.[11]

The figures in Table 42 indicate that, despite steady improvement, excess capacity continues to persist. The figures may be somewhat misleading. Some of the remaining excess capacity may result not from structural causes, but from the seasonal demand factor. We need to divide excess capacity into these two components. Since most consumers contract for coal requirements for a year or more, excess capacity resulting from seasonal fluctuations will have less adverse effect on prices than will persistent excess capacity.

We studied only the years 1956–1960, the most recent period covered in Table 42. For each state, tonnages for the six peak winter months (October through March) were projected on an annual basis.[12] The resulting potential production totals were compared with each state's capacity totals. Using this yardstick, the weighted average midwestern operating rates, in percentages, for 1956–1960 were: Indiana, 88.7; Illinois, 82.6; western Kentucky, 77.0; the midwestern average, 81.8.

Most mine operations are geared to handle a peak seasonal demand; hence, when idle capacity persists during this period, it presumably results from frictional forces. On the adjusted basis the cyclical or secular excess capacity in the midwest falls from 21.9 to 18.2 percent. (If excess capacity is based on unused

[11] The rate for the balance of the industry would be even lower — approximately 70.2 percent — if the midwestern share were eliminated.

[12] We adjusted tonnages for the other six months to allow for the ten-day miners' vacation in late June and early July. The October through March tonnage totals for Illinois and Indiana, predominantly unionized districts, were multiplied by 1.944 to project the figures to a twelve-month basis. A factor of 1.96 was applied to the western Kentucky figures, allowing for nonunion production during the vacation period.

capacity for the peak tonnage month, the average figure falls to approximately 13 percent.) Note that results for the three districts vary. A larger share of Indiana's excess capacity stems from the seasonality of demand than is true of western Kentucky. The adjusted Illinois figure lies between the figures for her sister districts. Two factors probably account for this pattern. First — and not too important — western Kentucky operators move a larger share of their domestic tonnages into the warmer southern states than do Illinois and Indiana shippers, thereby reducing their seasonal fluctuations. More importantly, western Kentucky moves coal during the summer months into the lake-cargo market. These shipments are large enough to offset the traditional summer slump, the result being that western Kentucky's adjusted excess-capacity figure coincides with the Table 42 figure, unadjusted for seasonal fluctuations.

We can dig deeper into the problem of excess capacity by subdividing the nonseasonal figure into its chronic or secular segment and the portion attributable to short-cycle fluctuations. To study the short-term movements we fitted a trend line to the 1950–1960 midwest tonnage data using the least-squares method. There was little evidence that overcapacity resulted from short-term changes in demand. In the last part of the period, when residuals lay above the trend line (expansionary periods), the operating rate differed only slightly from the 1956–1960 average level of 78.1 percent. In 1960, for example, when midwest output moved 3,500,000 tons above the eleven-year trend line, the operating rate stood at 80.3 percent, just 2.2 percentage points above the 1956–1960 average figure. Excess-capacity changes responding weakly to short-period shifts in output, we conclude that most of the remaining nonseasonal excess capacity is attributable to the secular component. Unless the postwar trend toward reduction in this secular portion abates, the future should witness even higher average operating rates in the midwest though they should soon bump against a ceiling.

Recent events have reduced whatever tendency toward cyclical excess capacity may have existed in the past. A study of the residuals around the regression line revealed the change. During the earlier years studied, the amplitude of the short-cycle changes were much greater than those encountered in recent years. Yearly output changes formerly ranged up to 10 to 15 percent, continuing a pattern of fluctuations that traditionally existed in the industry. Increasing reliance on the stable electric utility industry has reduced the amplitude of the fluctuations since 1955, average annual changes since then amounting only to 4.4 percent.

The significance of the 18.2 percent adjusted excess-capacity figure (100 percent − 81.8 percent), is fully understood only in light of the uninterrupted trend in the midwest toward reducing idle capacity. Taken alone, the figure indicates that competitive pressure can be expected from the redundant facilities seeking sales outlets. As long as this excess capacity hangs over the industry it is bound to have a depressing effect on prices and earnings; however, the forces described in Chapter VI modify this effect in several ways. Demand having shifted from retail to utility sales, the small producer, relying on domestic sales, has lost his position in the market. And his weak coal reserves position prevents any improvement in his situation. Furthermore, the oligopolistic control of desirable reserves keeps potential rivals from entering the industry and adding to excess capacity.

Existing excess capacity disappears gradually in three ways: (1) submarginal mines, sustaining losses for several years, finally go out of business; (2) marginal operators, having exhausted their coal reserves at existing mine sites, refuse to renew their investment in new mine properties; and, in rare cases, (3) mines involved in mergers are closed, production being shifted to lower-cost properties. Most of the reduction in midwestern excess capacity has stemmed from the first factor. High-cost mines, unable to withstand competitive pressure, have withdrawn from the industry. Many of them were geared to supply the relatively high-priced railroad and retail markets, and when those markets shrank they were ill-equipped to meet competition in the remaining utility and industrial coal markets.

This process of attrition, though somewhat unique, is not totally dissimilar from the exit pattern which previously existed in the midwestern industry. But excess capacity in the future should remain lower than it has been in the past because of a changed re-entry pattern. The practice of re-entry has never been widespread in this region in view of the restrictive topographical conditions. It is even less likely to exist in the future in light of the entry-deterrent forces delineated in Chapter VI.

Contrast conditions in the midwest with those in eastern mining districts where the removal of stubbornly persistent excess capacity is often impermanent. "Snowbird" mines, squeezed out of the industry during adversity under one management, re-enter the industry under new ownership. As long as reserves remain unmined, a small investment is sufficient to resume operations, adding to this region's chronic burden of excess capacity. Relieved of this problem, the midwest should continue the trend toward removal of its excess capacity with its attendant effect on improving efficiency and on easing the pressure on prices.

The level of excess capacity varies not only among coal districts but among mines within each district as well. There appears to be a fairly clear pattern of excess capacity based on mine and company size. In 1959, the seven largest Illinois operators, producing all but a small fraction of their tonnage from mines with an annual output exceeding 500,000 tons (and most from +1,000,-000-ton operations), worked their strip mines an average of 201.4 days, their deep mines 218.4 days. Illinois strip mines in the 50,000- to 500,000-ton class on the other hand, operated 181.5 days, while deep mines in the same class worked 184.4 days. The smallest mines — those in the 10,000- to 50,000-ton class, operated only 166.3 and 153.1 days for strip and underground, respectively.

What causes this working-time pattern? Working more days per year, the larger mines operate farther down their short-run, average-cost curves than do the smaller mines. But this condition represents a tenuous advantage since, by expanding output, the smaller firms could duplicate the larger firms' favorable positions. Thus the advantage lies elsewhere. It rests in the position of the larger firms' long-run cost curves. Possessed with superior reserves, firms operating the larger mines produce with long-run costs lower than their smaller competitors. And, being financially capable of securing the advantages of large-scale output, they can enhance their positions by moving to the right along their long-run cost curves. These twin advantages are reinforced by the larger firms' strong sales connections which help to give them improved running time. Presumably stronger finances would permit small operators to overcome two of the three disadvantages facing them. They could purchase both the necessary plant and sales talent to overcome these inadequacies, but they would be hard pressed to correct their weak coal reserve position in light of the reserve ownership pattern previously indicated.

We have found that the competitive sector of the industry, composed of the smaller operations, exhibits more excess capacity than does the oligopolistic sector. But is it not feasible to argue that the reverse condition might exist, i.e., that large producers, recognizing their mutual interdependence, would contract their output to increase profits and thereby increase excess capacity, unlike the competitive sector which would not curtail output? The existence of the competitive fringe makes this policy a dangerous one for the large firms. Though the character of market demand reduces the threat from the smaller firms, still these firms cannot be completely ignored. Their presence would render partially ineffective a conscious policy by the oligopolistic sector to curtail output. The small firms, with their heavy excess capacity, would stand ready to fill the gap left by output-contracting oligopolists.

Society stands both to gain and lose from the existence of less idle capacity in the larger than in the smaller mines. The trend being unmistakably toward large mines concentrated in fewer hands, the future should call for a larger percentage of production from mines with relatively low idle capacity, gained at the expense of the small operations which traditionally have burdened the industry with heavy excess capacity. This development should improve over-all technical efficiency. On the other hand, the economy *could* lose from the potentially stifling effects on price competition that a reduction in the number of sellers might create. The ever-present possibility of substituting alternate fuels, however, puts an effective ceiling on coal prices irrespective of the concentration level reached by the coal industry.

Extent of Optimum-Scale Production

In Chapter V we determined the approximate plant sizes required to achieve optimal efficiency. We recognized limitations in that analysis; we need now to reiterate those shortcomings and add an important additional one before estimating the percentage of midwestern output coming from these plants. To repeat: we use output per man-day rather than costs per se as an efficiency criterion. Differences in seam and mining conditions make comparisons among mines difficult. An entrepreneur cannot exactly duplicate costs achieved by an optimum-scale plant merely by duplicating its physical facilities if the underlying geological conditions differ appreciably. More severely limiting is the problem that achieved productivity levels (hence resultant costs) may not reflect obtainable scale economies. And the dynamic character of technological change constantly shifts the optimum-scale level, particularly in strip mining. These factors handicap the analysis and permit only a rough approximation of the extent of "optimum" scale production.

Using the data underlying Table 24, we computed the Illinois tonnage produced by strip operations producing over 30 tons per man-day and underground mines producing more than 15 tons per man-day. Production from plants producing at these efficiency levels totaled 32.9 million tons, 75.2 percent of Illinois's 43.8-million ton production. Although these efficiency standards fell slightly below average efficiency rates for the most productive mines, they were probably still sufficiently high to include plants operating at approximately optimum scale. The less efficient mines among the group were large operations with modern equipment and techniques which produced at efficiency levels far above the industry average. If their productivity fell below the level of their competitors it was attributable not to the inefficiencies of small scale but to the disadvantage of operating under less desirable mining conditions.

Most of the production from plants operating at the optimum scale came from strip mines producing over 500,000 tons annually and underground mines in the plus 50,000-ton class. The majority of the tonnage from mines operating at suboptimum efficiency levels came from small, local truck mines and marginal rail mines. These two inefficient elements of the industry for the most part constitute the competitive "fringe" surrounding the oligopolistic "core."

In the chapter on the industry's structure we noted the tendency toward reduction in the size of this "fringe." We might reasonably ask how it continues to survive at all in the face of its inferior productivity performance. Some of the inefficient small firms comprising the "fringe" exist by supplying coal to geographically isolated markets in areas adjacent to the mines. Moreover, the persistence of the "fringe" may be somewhat misleading because membership in the group is not necessarily constant. New small firms, whose expectations exceed their capabilities, enter the industry from time to time only to languish after sustaining losses for several years. Their places are subsequently taken by other hopeful newcomers.

Two additional groups comprise the persistent competitive "fringe." First is the small firm run by an owner-manager or by a family group. This type of firm often continues to exist despite its inefficiency by sacrificing a fair rate of return for the independence of entrepreneurship. The second group hangs on by cutting wages instead of adopting efficient techniques. In Indiana and in all but a tiny pocket in southern Illinois where small non-union mines flourish, allegiance to the U.M.W.A. is strong enough to prevent wage cutting. Less highly organized western Kentucky presents a different picture, however, with approximately 15 percent of that district's tonnage still coming from nonunion mines.

It is difficult to determine the extent to which the inefficiencies of unduly small scale are eradicable. As long as firms remain small, they can do little to overcome the managerial and financial deficiencies that contribute to their inefficiency. The question of eradicability thus hinges largely on whether or not unduly small firms will persist in the industry. The lure of owning one's own business will continue to attract some marginal firms. They will be joined by those seeking to carve out a niche in isolated geographical markets. (In many cases, of course, individual entrepreneurs will fall in both categories.) The rapid drying up of the retail market and the insistence upon adequate reserves by many large buyers, however, reduces the potential size of these two groups. The substitution of wage cutting in place of the adoption of efficient mining techniques might well continue in the immediate future throughout parts of western Kentucky. The tendency could be weakened either through the use of a strong organizing drive by the U.M.W.A., from an amendment in T.V.A.'s buying policies outlawing nonunion coal, or as a result of tighter control over reserves by existing unionized companies. Only the third alternative appears feasible in the near future, and if it is used it should be only partly successful.

Effect of Mergers on Productivity

What is the effect, if any, of merger activity on productivity rates? To answer this question we compared the productivity records of merged mines with those of a random sample of mines operated by companies not involved in mergers.[13] Average annual productivity increases for Illinois and western Kentucky mines merged between 1950 and 1959 amounted to 4.03 percent; output per man-day in the sample group rose 3.50 percent. Differences between strip and deep mines were striking. Productivity for merged strip mines climbed an average of 9.16 percent annually while the sample mines' productivity fell 0.50 percent. Comparable figures for the deep mines were 1.62 and 5.60 percent.

The differences depended in part on the level of productivity in the base year. Most of the deep-mine mergers involved mines already operating at efficient rates, the room for improvement therefore being slight.[14] Many of the sample group mines moved from low to high levels of mechanization, bringing them up to

---

[13] We studied productivity records for Illinois and western Kentucky mines operated by the following coal companies involved in mergers for the period 1950 to 1959: Peabody, Truax–Traer, Ayrshire, West Kentucky, Ziegler, and Freeman. Productivity records for Indiana mines were unavailable. We selected for analysis only those merged mines which were still in operation in 1959. Annual average (not discounted) productivity increases were computed for each mine, comparing productivity rates in 1959 against those achieved in the year immediately prior to merger.

The sample of nonmerged mines was constructed as follows. We selected in alphabetical order a group of mines from the Kentucky and Illinois Departments of Mines and Minerals' Annual Reports to make up a composite group that approximated the characteristics of the mines in the merged group. The list included the same ratio of strip to deep tonnage (2:1) as was found in the merged group, and productivity changes were measured for the same time periods in approximately the same proportions.

[14] The merged group's average failed to rise more because of the depressing influence of West Kentucky Coal Company's dismal record. Its productivity rate actually declined as a result of mismanagement.

levels previously gained by the merged mines. The spectacular increase in the merged strip-mines' productivity rates came from the widespread use of more capital-intensive mining methods introduced by the surviving companies. In the absence of the introduction of these techniques in the sample mines, their productivity rates declined with age. The extent to which future mergers improve efficiency should depend on the level of productivity existing prior to the mergers. The ceiling on scale economies in deep mining limits the potential benefits to be derived from deep-mine mergers, but the possibilities for improvement in strip efficiency from mergers are apparently large if the past is an accurate gauge.

Income Distribution

The previous discussion of midwestern coal earnings studied the rewards going to one of the factors of production. We broaden the analysis now to include the other factors, looking particularly at returns to mineral owners. Whenever possible we measure comparative returns; however, most of the analysis perforce must be impressionistic.

The effect of an oligopolistic market structure on the distribution of income among owners of capital and the other factors depends partly on the extent of collusion existing among the oligopolists and the effectiveness of the industry's entry barriers. These factors influence the degree to which the firms collectively can act to approximate the monopolist's price-output decisions. Adopting the posture of a monopolist, the industry enhances its profits, bringing about an inequality of income distribution. But labor, backed by a strong bargaining position, can redress the inequality. A strong union can exert pressure on management to share with it the fruits of the industry's monopoly position.

Conditions surrounding returns to land or mineral owners differ somewhat from those facing other factors. Location and quality of reserves, their relative scarcity, ownership patterns, and imperfections in the coal reserves market all contribute to the share of income distributed to this factor.

Central to this discussion is knowledge of the form ownership can take. Mineral ownership in underground operations occurs in one of three ways. The mineral rights may be owned outright by the mining company, permitting it to extract coal from under the surface owners' lands subject to deed and/or statutory restrictions limiting the mining area to prevent subsidence. The mineral expense per ton for the company is the purchase price of the mineral acquired divided by the merchantable tons extracted from the property. Alternatively, a coal and landholding subsidiary may acquire mineral rights and in turn may lease coal to the parent company on a stipulated royalty basis. The lease arrangement may favor the subsidiary, transferring profits to it from the parent company. Finally, mineral ownership may reside with outsiders—either individuals or nonaffiliated coal and land companies—who lease the mineral to the operating company.

Ownership is similar in strip mining, with a major exception. There, the need to disturb the surface to recover the buried mineral complicates matters since title to the surface land and subsurface minerals may rest with two parties rather than with one. Thus, for example, a landowner can receive a royalty for each ton of coal extracted from beneath his land though he owns only the surface. Other parties may own the mineral rights, and they too may earn royalties in payment for the removal of their coal. Usually, however, surface and mineral is owned in fee either by the operating company or by a lessor.

If the operator leases his reserves from mineral owners the lease agreement stipulates a royalty rate expressed as a fixed percentage of the average mine realization or as a flat cents-per-ton figure. The rigidity of the return to the land or mineral owner from the incorporation of the royalty rate in a lease agreement has interesting income distribution implications. Fixing the rate determines the income share accruing to the landowners irrespective of the shares earned by other factors, provided the tonnage extracted remains unchanged. Altering the output rate affects royalty earnings during a given time period, but it cannot influence the total return earned by the mineral owner over the life of the reserve. And even annual royalty earnings can be protected from the effect of tonnage reductions if the agreement includes a high guaranteed minimum-royalty payment provision. A sophisticated lessor usually demands this protective feature and, additionally, insists that the guarantee take effect upon the signing of the agreement to discourage the operator from putting the reserve into "mothballs."

Having a relatively fixed return, the lessor is powerless to manipulate his relative income share. This condition can work for or against him depending on the circumstances in each case. Generally he gains relatively during periods of declining demand and suffers when demand rises. (This generalization implies that the direction of wage and price movements corresponds to the direction of demand changes.) A fixed royalty rate represents a relative improvement in income distribution when wages, prices, and profits are falling. Conversely, when demand increases, the landowner's relative position generally weakens. His static royalty rate contrasts with normally increasing wage and profit rates which shift distributive payments toward labor and capital.

Two factors can alter these conclusions. As indicated, in the absence of a guaranteed minimum-royalty proviso, the lower output flowing from reduced demand reduces income to mineral owners.[15] Under this condition landowners probably gain at least relative to owners of capital. Even at low tonnage rates mineral owners receive payments long after returns to capital become negative — a prevalent condition in this storm-tossed industry. The occasional use of a royalty rate geared to the mine's average realization may influence our generalization though its effect is probably minimal. Thus a price reduction from $4.00 to $3.60 per ton, reducing a 5 percent royalty payment from 20 cents to 18 cents, probably has a proportionally more severe effect on earnings, especially if the reduction stems from downward demand pressure and not from the effect of improved productivity. Changes in the distributive shares to landowners vis-à-vis wage earners is more difficult to gauge and would depend largely on labor's bargaining power.

If the operator owns his coal reserves (either outright or through a landowning subsidiary) rather than leasing them, the distinction between distributive shares to land and capital loses

---

[15] We assume here that reduced demand spells lower output. It might, of course, call for the same output to be sold at lower prices, or a combination of lower prices and reduced output.

some of its meaning since ownership of both assets is common. Nevertheless, a distinction is still necessary. What may at first appear to be a return to capital in mining is really a composite made up of separate returns to capital and land (i.e., mineral resources).

As with land, mineral resources yield rent because of (1) their superior quality or (2) their superior location. The first factor is responsible for different values being placed on reserves in the same general location while the second accounts for different values for identical coal beds in different locations. Capital, labor, and entrepreneurship, of course, can earn the same two kinds of rent. But the second of these—locational rent—can be eliminated through transfer of these movable factors to the favored location. Again, as with land, the immobility of buried coal reserves prevents a similar transfer. Unlike land, however, mined coal becomes mobile. Thus the locational rent earned by a favorably situated coal reserve depends on the transportation-cost advantage it holds over competitive reserves.

Swiftly moving external forces can create quality or locational rent for coal reserves virtually overnight, the windfall increasing mineral owners' distributive shares. Two fairly recent developments in western Kentucky support the point. Until the early 1950's most western Kentucky strip operators wasted the no. 12 seam, an inherently dirty seam overlying the more desirable no. 11 coal. Washing the coal, though improving its quality, failed to make it marketable for discriminating buyers. However, the installation by T.V.A. of burning equipment capable of handling this off-grade fuel transformed it from a nuisance into an acceptable product. Operators whose reserves contain both no. 12 and no. 11 coal seams mine the former virtually as a by-product of mining the latter, incurring little additional expense and reaping a substantial rent.

The second development involved deepening the channel and widening the locks in the Green River, located in Kentucky, to open the river to the transportation of large barges from western Kentucky mines. Coal-bearing land, adjacent to the river but previously inaccessible to cheap water transportation, rose sharply in value. Land bearing eleven feet of coal in two seams produced royalty earnings of up to $3,500 per acre based on 14,000 recoverable tons per acre at a royalty rate of 25 cents per ton. This inflated land value contrasted sharply with its value of $100 or so in its alternative use as farm land. Since national wage bargaining determines wage rates, labor locally was unable to press effectively to share the fruits of locational advantage. Income distribution tipped in favor of the leasing landowner or the coal company depending on the circumstances in each case.

One of these circumstances — greatly influencing mineral owners' distributive shares — is the degree of market perfection in the coal land market. The reader might reasonably wonder whether excess profits stemming from superior location and quality would not automatically accrue to landowners as rent. Custom and market imperfections may prevent it. J. V. Nef noted forces at work three to four centuries ago denying landowners the full fruits of their mineral monopoly market position:

Although, in the past, the landlord sometimes succeeded in claiming a portion of the return which arose from advantages of situations or of fertility, the royalties paid at different mines have never been an accurate measure of the differences in natural advantages between these mines. Mining conditions and the state of the market for coal changed from year to year, and even from month to month, while colliery leases have always been drawn for much longer periods, during which the royalties have been fixed, in the sixteenth and seventeenth centuries most commonly at a definite sum per annum, and since the seventeenth century more frequently at a definite sum per unit of coal extracted. No method of assessing royalties to absorb everything above the usual profits of stock throughout the term which leases have to run has yet been devised.[16]

In addition to the situation described by Nef, unequal market knowledge and bargaining power prevent coal royalty rates from attaining their "proper" values. The parties to a royalty bargain are an experienced, knowledgeable coal operator on one side and either an equally well-informed coal operator or coal landowner or an individual landowner (usually a farmer) on the other. Experienced coal operators command a significant bargaining advantage over the uninitiated independent landowner. First, the operator is an experienced negotiator, aware of relative coal land

[16] J. V. Nef, *The Rise of the British Coal Industry* (London: G. Routledge & Sons, Ltd., 1932), vol. I, 329.

values. Moreover, he knows better than the farmer how the land parcel fits into the mosaic of landholdings that comprise a coal field. The operator's possession of detailed drilling records, generally unavailable to the landowner, strengthens his position. This inequality in market knowledge and bargaining power shows up in the negotiated royalty rates. The operator's strong position may win him rates which fall far short of those which truly reflect the landowner's differential advantage.

Conditions are different when the operator deals with another operator or with a coal landholding company. Cognizant of coal land values, an experienced lessor demands royalties that reflect quality, location, and scarcity values. In periods of rising coal prices he is apt to receive royalty rates that fluctuate with coal prices as an inflationary hedge. Or he may forego royalties and receive instead stock from the operating company, capitalizing on his land's differential advantage by sharing the operator's profits. Duncan Coal Company made such an arrangement with Peabody Coal Company when it relinquished its desirable western Kentucky strip coal acreage near the Green River in exchange for Peabody stock. We have record of a similar arrangement undertaken in Indiana over thirty years ago with substantial rewards to the mineral owner. In this instance the landowner contributed 600 acres of coal land plus $25,000 for a 50 percent stock interest in a new strip-coal venture; his partner, who was to operate the mine, invested $200,000. The landowner, a veteran coal operator, earned over $3 million in dividends and increased equity over the mine's life, an average of 63 cents per ton for the tonnage originally contributed to the venture. Subsequently the firm leased additional acreages of equally desirable land in the same area, mostly from small landowners at a royalty rate of 10 cents per ton.

Similar cases, though, are rare in the midwest. This condition is a function of reserve ownership patterns existing there. Most of the midwestern reserves, not controlled by existing producers, are owned by the individual landowners who control the surface above the coal. In the Appalachian region, however, large coal and landholding companies (many of them railroad company subsidiaries) control extensive coal acreages. And their royalty rates tend more to reflect a reserve's differential advantage than they do in the midwest with its diffused coal ownership structure.[17]

Another factor influencing income distribution is the relation between productivity changes and wages paid to labor, particularly in this industry in which labor costs loom large in the product's total costs. Figure 10 plots this relation and the trend in midwestern coal prices, for the period 1950–1960. A weighted average tons-per-man-day figure represents productivity, with yearly tonnages for each state serving as the weights. The average daily wage index is a composite of the average daily wage for Illinois coal mines and wage rates for the modal wage classifications in Indiana and western Kentucky, the yearly figures again weighted by the states' annual tonnages.

Expressed in relatives, the tons-per-man-day figure has consistently remained ahead of the average wage line. In the absence of total industry earnings' figures for the period we are unable to determine the effect of this relationship on distributive shares. The dip in the average price line, however, coinciding with the start of the big spread between productivity and wages, leads to the inference that the consumer gained at the expense of the workers at least in the years up to 1955. If the earnings' figures in Table 40 depict fairly the trend in midwestern earnings generally, the consumer gained at the operators' expense as well. But the earnings' decline would have been much more severe than it was if productivity had failed to increase as rapidly (or if the wage level had lagged as badly) as it did. Between 1950 and 1955 weighted average productivity in the midwest increased 66 percent while wages rose only 25 percent. In the period 1955–1960, however, labor redressed the imbalance, increasing wages 31 percent

[17] This statement does not contradict the previous assertion in the chapter on entry barriers that the existing companies possess the most desirable reserves, though it may appear to do so. What we have described are the conditions under which many of the existing operators' reserves were acquired. Moreover, adjacent to workable blocks of coal are small, coal-bearing acreages which an operator acquires to round out his holdings. Additionally, despite the trend toward seller concentration in the midwest, there are still small- and medium-sized operators who require coal reserves to continue operations. The coal land market described here applies in these instances as well.

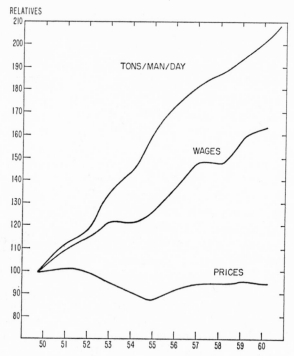

FIGURE 10. Wages, prices, and productivity, midwestern mines, 1950-1960

against a 26 percent productivity gain. The improvement in operators' earnings stemmed from a general improvement in the price level which turned upward in 1956.

Conservation of Natural Resources

The midwest's greater use of the strip-mining technique improves its performance in the conservation of coal resources. One hundred percent recovery of coal reserves using any technique is infeasible. If either strip- or underground-mined coal is washed, washery losses reduce coal recovery 5 to 7 percent below rates attained in raw coal operations. Additional strip-coal reserves are lost from spillage, from slides covering the coal seam, and from other pit losses. But losses from underground mining are even greater. Large quantities of coal are lost in abandoned support pillars left by underground mines using the room-and-pillar mining system.[18] Additional underground-mining practices reduce recovery. Coal is often abandoned when faults are encountered; and the danger of subsidence prevents the removal of coal under towns, roads, and rivers. These deterrents to resource conservation result in an approximately 30 percent greater coal loss in underground mining than in strip mining. Most estimates place strip-coal recovery at 80 to 90 percent. Gilbert H. Cady indicates recovery of underground-mined no. 6 coal in Franklin County, Illinois, to be around 50 percent.[19] A broader Bureau of Mines' study of eleven Appalachian region counties reveals recovery rates of 46.2 to 62.1 percent, the average being 53.7 percent.[20]

There are several additional elements in the strip-underground equation of resource conservation. Strip mining increases coal recovery slightly by mining coal too shallow to be recovered at all by deep-mining methods. On the other hand, some strip operations lead to grossly wasteful practices. In the absence of an authority insuring that coal resources are wisely exploited, strip operators are free to uncover coal to any depth they desire. Many small operators carelessly skim off the coal under light cover, leaving unmined reserves that are too deep to be mined with their limited equipment. Noting that the "low-ratio" coal has been removed, large operators are discouraged from entering the field, leaving the deep coal untouched, wasted.

A final conservation consideration engenders bitter controversy among various economic interests in the mining fields. This consideration broadens the study of conservation, asking not what method of mining best conserves coal resources, but rather questioning how all resources — surface and subsurface — can best be utilized. Are there disadvantages to the use of strip mining that outweigh its coal-conserving advantages?

Opponents of strip mining argue that the destruction of farm lands and creation of aesthetically unpleasant landscapes caused

[18] Schurr and Netschert estimate 90 percent of underground coal comes from mines using this system in *Energy in the American Economy, 1850–1975*, p. 312.
[19] Averitt, *Coal Resources of the United States*, p. 12.
[20] *Ibid.*, p. 13.

by stripping vitiate whatever advantages this mining technique may otherwise possess. Its defenders point to the cost-saving and coal-conserving characteristics of strip mining and cap the argument by stressing that only a small percentage of farm land is adversely affected. Strip coal land, it is argued, accounts for only .302 percent of Illinois's 36 million acres.[21]

The economic justification of stripping as opposed to using the overlying farm land for agricultural purposes is deceptively convincing. Herman D. Graham, in a study of Illinois's stripping, argues as follows: In 1940 the average gross income of farm land in Fulton and Knox Counties, Illinois (two prominent strip-mining counties) was $18.11 and $22.35 per acre respectively.[22] In the same year the average gross income (defined as the f.o.b. sales realization) per acre from strip-coal mining was $5,193 in Fulton County and $3,461 in Knox County. Thus Graham argues "it would take 155 years for the value of Knox County crops and 288 years for that of Fulton County crops to equal the value of coal extracted per acre."[23] This reasoning is specious. If this coal had been less efficiently mined, its cost (and price) would have been even higher, presumably increasing welfare still more!

Thus we return to the central issue: the creation of aesthetically unattractive land areas. Responding to this condition, the states of Illinois, Indiana, and Kentucky (among others) have enacted laws regulating the use of stripped-over land. Fulfilling the laws' requirements reduces slightly strip-mines' cost advantages by entailing expenditures by strip operators to reclaim the disturbed land. The three states' laws are similar though Illinois's and Indiana's are somewhat more restrictive than Kentucky's. The first two states give the operator the option of several reclamation possibilities, the alternative to exercising an option being the forfeiture of a bond which costs more than the expense of reclamation.[24]

The costs of land reclamation are somewhat obscured by the

[21] Statement of Midwest Coal Producers Institute, Inc., Jan. 1, 1963.

[22] *Ibid.*, pp. 46, 47.

[23] *Ibid.*, p. 47.

[24] We are grateful to Mr. L. E. Sawyer and Mr. L. S. Weber who perform land conservation functions for the Indiana and Illinois coal industries for the data used in this section.

operators' practice of burying these costs with mine operating costs, a practice which reduces their income tax liability by reducing the capitalized value of the reclaimed land. Nevertheless, the state coal associations have gathered cost figures over the years which are probably reliable, though perhaps incomplete. Operators usually afforest reclaimed land, seed it for pasture, striking off the spoil-bank tops prior to planting, or grade the land for crop cultivation, preparing the land "in such a manner that the area can be traversed with farm machinery reasonably necessary for such use." [25] Costs vary a great deal for each of these operations. The Indiana Coal Association spent an average of $19 per acre to afforest Indiana strip land between 1945 and 1962. Illinois has spent an average of $30 per acre, with a range of $18 to $50 per acre. To seed the land for pasture has cost Illinois strip operators an average of $40 an acre, the range being $25 to $75. Complete grading has cost $100 to over $300, with an average of $150 per acre.

For an acre originally underlaid with five feet of coal these Illinois averages represent expenditures of four mills to two cents per ton of mined coal. Returns to the operator presumably depend largely on the expenditures he makes. Expenses of afforestation are slight but the discounted returns are relatively minor, also, compared to the immediate returns from farming graded land. Usually the character of the overburden determines whether the operator afforests or levels the stripped-over land. In any event since most spoil banks can be seeded for pasture or planted with seedlings, the operator willing to do no more than comply with the reclamation laws can fulfill his obligation at little expense. And with proper land management and favorable soil conditions he may convert a wasteland into a profitable side venture.

[25] The Open Cut Land Reclamation Act, Sect. 6 (f).

CHAPTER IX

Final Appraisal

There is a conflict in the literature concerning the degree of competition in the American bituminous coal industry. Interestingly, two editions of *The Structure of American Industry* present two contrasting views that mark the limits of the divided positions on this question. Writing in an early edition of the volume, Norman Leonard cites:

(1) The existence of over 6000 sellers in 1946, no one of which controlled more than three percent of total industry output plus the fact that the industry has no price leader.

(2) The absence of both patent restrictions and grants of exclusive rights that tend to impede entry.

(3) The inability of firms to differentiate their products even though coals inherently differ from one another. He asserts that the ability to translate coal values into a cost per million B.t.u. gives it a quality of homogeneity. Furthermore, he argues that consumers indicate no special preferences for coal from individual firms.[1]

Leonard's arguments are weak on nearly all counts. He ignores the competition-reducing forces of distance and high-transportation charges that divide the country into regional submarkets. His conclusions on entry conditions are partly correct, but are supported by different arguments from those he marshals. In large segments of the industry, entry remains easy, not from the absence of patent restrictions and exclusive rights arrangements, but

[1] Norman H. Leonard, "The Bituminous Coal Industry," *The Structure of American Industry,* ed. Walter Adams (New York: The Macmillan Company, 1950), pp. 34–35.

because there are fairly low capital requirements and accessible coal reserves. Changing market demands and a trend toward larger mines and greater mechanization requiring larger capital outlays are modifying this characteristic in much of the industry.

The conclusion that coal is homogeneous is equally unsound. The fact that using a cost-per-million-B.t.u. approach converts different coals to a common evaluation base should not obscure the existence of differences in ash, moisture, sulphur, free swelling indices, etc., that affect coal's value. These factors, plus individual companies' sales efforts, lead to consumer preferences for specific coals, Leonard's assertion to the contrary notwithstanding.

Armed with virtually the same information, Jacob Schmookler, in a revised edition of *The Structure of American Industry,* draws a different conclusion. "Unfortunately, the basic premise that coal is, or ever was, a purely competitive industry is unsound," Schmookler asserts.[2] On the selling side he notes the existence of product differences, negatively sloped individual firm demand curves, selling expenses, locational advantages, price discrimination practices, and individual consumer preferences—all of which lead to imperfectly competitive markets. Professor Christenson's recent study leans to the Schmookler view.[3]

Parker, on the other hand, studying the social implications of economic conditions in coal, sees evidence of both monopoly and competition in the coal industry, but he bases his conclusion on a narrow interpretation of the facts.[4] He asks only: On balance do industry conditions lead toward or away from "combinations" of selling units? He notes as forces tending toward monopoly: the fact that the industry operates under increasing returns as capacity output is approached, the inelastic demand for coal, transportation rivalries "that bring monopoly through intensified competition," [5] and "geographical compactness." Parker's path to monopoly is circuitous. The foregoing forces "all immediately make for wolfish competition," the antithesis of monopoly, but

[2] Jacob Schmookler, "The Bituminous Coal Industry," *The Structure of American Industry,* ed. Walter Adams (rev. ed.; New York: The Macmillan Company, 1954), p. 79.

[3] Christenson, *Economic Redevelopment of Bituminous Coal.*

[4] Parker, *The Coal Industry,* pp. 5–6.

[5] *Ibid.,* p. 6.

eventually there is an "obvious desire on the part of enterprise to eliminate losses through combination." [6] Forces making for competition include: the wide distribution of land ownership, the difficulty of valuing properties easily, ease of entry, and legal restraints. These factors are monopoly-deterrents, preventing firms from combining through informal arrangements or the use of mergers or consolidations.

Most of the apparent confusion over this question stems from two factors mentioned at the outset of this study. First is the obsoleteness of some of the studies written before conditions changed in the industry. The coal industry of the 1960's differs from the coal industry of the turbulent Guffey Act days. Second, distortion arises from generalizing about an industry composed of several diverse parts. This study has sought to correct both deficiencies by taking a current view of one of the industry's major segments.

Our findings in the midwest support Schmookler's conclusions. This part of the industry qualifies as imperfectly competitive on three major counts: numbers of sellers, entry conditions, and character of product.

Though several hundred sellers still remain in the midwest, control of most of this region's output rests in relatively few hands. The industry can be accurately labeled an "oligopoly with a competitive fringe." The trend toward few sellers stems partly from attrition among the fringe which depends strongly on declining consumer market segments. But much of the increased concentration derives from a conscious policy of growth through mergers conducted by most of the remaining large firms. Is it fair to question the motives behind these mergers? Is it not reasonable to suspect that the mergers attempt to reduce the competitive pressures burdening an industry long wracked by depressed prices that flow from unstable demand conditions and oppressive excess capacity? This is a possibility, though we have no evidence that this motive lies behind the merger movement.

A more likely explanation is the desire of firms to grow merely to meet the modern needs of the industry. The industry's salvation depending on its participation in the expanding electric

[6] *Ibid.,* p. 6.

utility market, the successful operator must gear himself to meet its demands. And its demands call for large size, for large reserves strategically located to satisfy the needs of utilities operating in different market areas, and for large, capital-intensive mining operations to guarantee low-cost operations which are large enough to permit taking advantage of cost-saving modes of transportation.

The positions of the major firms, strengthened by the trend toward concentration, are strengthened further by the erection of moderate entry barriers. Again, utility requirements partly determine this condition. The premium on size penalizes the remaining existing firms and impedes entry for aspiring small outsiders. Add to this the concentration of the most desirable reserves in the hands of the existing large firms, and the barriers rise another notch. Still they are not insurmountable. The requirements of large size deter the traditionally small operator, but the $10-$20-million investment required to finance a 2-million-ton annual capacity plant is within reach of several large operators producing outside the midwest and of scores of firms from outside the industry who might seek entry. Overcoming the tight control of the "best" reserves might prove more formidable than surmounting the size barrier, yet two possibilities here exist. First, shifting locational requirements of the fast-moving electric utility industry may convert poorly located, mediocre coal reserves into very desirable assets. Additionally, major technological breakthroughs in underground mining may create the same effect by overcoming strip-mining's sizable inherent cost advantage.

Further buttressing the midwestern industry's position is its heavy reliance on the electric utility industry. The stability and growth of this major industry — further stabilized by the practice of buying much of its coal requirements on a long-term basis — rubs off on the midwestern industry. It would be convenient to argue that some of the structural conditions we have described (growth in concentration, entry conditions, etc.) explain the stability in midwestern coal earnings in recent years. Doubtless, these have been contributing factors. Certainly, however, the finger points as well to the fortunate partnership with the utilities.

The existence of market imperfections highlighted throughout much of this study does not mean that harsh, "excessive" competition cannot prevail in the industry. These "excesses" can have effects as undesirable as the "excesses" of monopoly. Indeed, much of the coal literature decries the adverse social and economic consequences of coal's "destructively" competitive pattern that so often has called for remedial governmental action in the past.

What are the symptoms of "destructive" competition? Maladjusted industries suffering the ill-effects of "excessive" competition exhibit several common tendencies, among them the following:

(1) Selling prices are consistently driven down to low levels over a long period of time.
(2) The depressed prices lead to chronically sub-par rates of return which are either negative or too low to cover interest costs of invested capital.
(3) Chronically submarginal returns to productive factors match entrepreneurs' subnormal earnings.
(4) Redundant resources reluctantly leave the industry.
(5) Returns to capital and labor fluctuate with greater amplitude in these industries than they do in others.[7]

The repercussions from destructive competition are severe. There is a continual misallocation of resources as surplus resources hang on like grim death long after their subnormal returns dictate that they move to new uses. The misallocation of resources leads to unequal income distribution, the returns both to labor and capital in the distressed industries being lower than those earned elsewhere. Finally, chronic depression leads to conservation wastes and to the use of inefficient, productive techniques, the latter being a function of the submarginal rates of return that prevent a "plowback" of earnings into new plant and equipment.

Is it fair to paint the midwest with the brush of "destructive" competition? The persistent view that this term aptly describes the industry demands that we answer the question.

On balance, it fails to exhibit either the symptoms or effects of

[7] Bain, *Industrial Organization*, p. 431.

a chronically depressed industry. In recent years, at least, prices have been fairly stable, as have rates of return on invested capital. The presence of a strong union, strengthened by its adherence to a policy of encouraging mechanization, has had the beneficial effect of accelerating the adoption of technological improvements. Only with respect to the resource allocation question does the midwest qualify as "destructively" competitive. Its inability completely to pry loose redundant resources prevents its deviating fully from the "destructively" competitive standards.

Though the midwestern industry escapes the perils of chronic depression, we certainly have no evidence that it is subject to monopoly's excesses. If rates of return exceed those earned in the balance of the industry, they are by no means abnormally high. If market dominance is tighter than it formerly was in this part of the industry or presently is elsewhere, there is still ample pressure from rival fuels and coal producers to insure effective competition. Though entry barriers rise, they are not insurmountable. This segment of the industry stands astride a boundary dividing competition and monopoly, with a foot in each camp.

If, as it appears, the "reform" in this region is permanent, the economy should benefit. The gains should be even greater if, as we also suspect, developments in the midwest presage a similar pattern in the balance of the industry.

Appendix A

The data in this appendix support the derivation of the hypothetical market boundary line in Figure 4. The line encloses a region within which at least one midwestern mining district commands a delivered-cost advantage over the lowest-cost outside producing district. Representative midwestern and outside coals used in the analysis are listed in Table 43 together with their assumed B.t.u. contents and f.o.b. mine prices.

TABLE 43. Assumed analyses and f.o.b. mine prices for representative midwestern and outside coal districts

District	B.t.u./pound	MM B.t.u./ton	F.o.b. mine price[a]
Western Kentucky (No. 11)	12,150	24.3	$3.50
Princeton–Ayrshire (Indiana No. 5)	11,500	23.0	3.90
S. Illinois (No. 6)	12,200	24.4	3.90
Belleville (Illinois No. 6)	11,000	22.2	3.75
Fulton–Peoria (Illinois No. 5)	10,700	21.4	4.05
Ohio No. 8 (Georgetown)	13,000	26.0	4.20
District 8 (W. Va., Sewell) ⎫ Inner	13,800	27.6	4.25
District 8 (E. Kentucky) ⎬ Crescent	13,600	27.2	4.25
Missouri (Northern)	11,000	22.0	4.00
Iowa	9,500	19.0	4.00
Jellico (Tennessee)	13,800	27.6	4.25

[a] Effective June 1, 1960.
Source: Confidential industry sources.

Delivered-cost computations were made for shipments from the different producing areas to eighty-six consuming points in the nine states included within and surrounding the midwestern market area. If a midwestern district enjoyed the lowest delivered cost per million B.t.u. to a particular consuming point, that point was included in the midwestern market area. The midwest's advantage diminishes naturally as shipments move farther away from the midwestern producing districts. At some breaking point delivered costs for the cheapest midwestern coal and outside district coal are equal. It was the essential task in this phase of the study to find the location of the various breaking points. These points were then connected with an unbroken line demarcating the hypothetical market area.

In most cases rail freight rates were used in computing delivered costs. To the major Lake Michigan and Lake Superior ports, however, we used combined rail-water rates. To inland cities the lower of all-rail and rail-water-rail rates were used. Table 44 lists the consuming points for which delivered costs were computed, showing both the transportation charges and delivered costs per million B.t.u. for the appropriate midwestern and outside districts. Included in the rail-water rates are transfer charges of eighteen cents per ton for midwestern coal moving over the dock of the Rail-to-Water Transfer Corporation at South Chicago. The going published and negotiated vessel rates from Toledo and South Chicago to the various lake ports were also used. A dock handling charge of sixty-seven cents per ton was included in the total charges for shipments reloaded at upper lake ports for further rail movement inland.

TABLE 44. Transportation charges per ton and total delivered costs per million B.t.u. for midwestern and outside district coal shipments to representative destinations

Wisconsin

To \ From	W. Kentucky Freight	W. Kentucky Delv'd. cost	Ohio No. 8 Freight	Ohio No. 8 Delv'd. cost	Inner Crescent Freight	Inner Crescent Delv'd. cost
Neenah	$4.77	34.0¢	$5.00	35.4¢	$5.41	35.0¢
Appleton	4.77	34.0	5.00	35.4	5.41	35.0
Kimberly	4.77	34.0	4.90	35.0	5.31	34.6
Antigo	5.28	36.1	6.15	39.9	6.53	39.8
Shawano	5.47	36.9	5.35	36.7	5.73	36.2
Nekoosa	4.77	34.0	5.96	39.1	6.34	38.3
Wis. Rapids	4.77	34.0	5.96	39.1	6.34	38.3
Stevens Point	4.77	34.0	5.96	39.1	6.34	38.3
Rothschild	5.28	36.1	6.13	39.8	6.51	39.1
Tomahawk	5.47	36.9	6.22	40.2	6.60	39.3
Milwaukee	4.05	31.1	3.85	31.0	4.28	31.0
Sheboygan	4.05	31.1	3.78	30.7	4.21	30.6
Green Bay	4.17	31.5	3.78	30.7	4.21	30.6
Oak Creek	3.69	29.5	3.85	31.0	4.28	31.0
Port Washington	3.69	29.5	3.85	31.0	4.28	31.0
Ashland	4.12	31.3	3.71	30.4	4.14	30.4

Michigan

To \ From	W. Kentucky Frt.	W. Kentucky Delv'd. cost	S. Illinois Frt.	S. Illinois Delv'd. cost	Ohio No. 8 Frt.	Ohio No. 8 Delv'd. cost	Inner Crescent Frt.	Inner Crescent Delv'd. cost
Niles	$4.90	34.6¢	$4.53	34.5¢	$4.77	34.5¢	$5.27	34.5¢
Three Rivers	5.09	35.3	4.74	35.4	4.62	33.9	5.12	33.9
Coldwater	5.15	35.6	4.80	35.7	4.22	32.4	4.72	32.5
Battle Creek	5.15	35.6	4.80	35.7	4.62	33.9	5.12	33.9
Kalamazoo	5.09	35.3	4.74	35.4	4.62	33.9	5.12	33.9
Grand Rapids	5.15	35.6	4.80	35.7	4.77	34.5	5.27	34.5
Jackson	5.40	36.6	5.05	36.7	4.22	32.4	4.72	32.5
Lansing	5.40	36.6	5.05	36.7	4.42	33.2	4.92	33.2
Ludington	4.23	31.8	—	—	4.29	32.6	4.72	32.5
Muskegon	4.17	31.5	—	—	4.31	32.7	4.74	32.6
St. Joseph	4.13	31.3	—	—	4.43	33.2	4.86	33.0
Petosky	4.64	33.5	—	—	4.21	32.4	4.64	32.2

TABLE 44 *(continued)*

Minnesota

From To	W. Kentucky		Missouri		Ohio		Inner Crescent	
	Frt.	Delv'd. cost	Frt.	Delv'd. cost	Frt.	Delv'd. cost	Frt.	Delv'd. cost
Winona	$5.77	*38.1¢*	$5.33	*42.4¢*	—	—	$7.10	*41.1¢*
Granite Falls	6.35	*40.5*	4.90	*40.5*	—	—	7.41	42.3
Brainerd	6.83	42.5	5.95	45.2	—	—	6.40	*38.6*
Fergus Falls	7.06	43.4	6.17	46.2	—	—	7.32	*41.9*
Worthington	5.87	*38.5*	4.78	39.9	—	—	7.22	41.5
Willmar	5.85	*38.4*	5.65	43.9	—	—	7.42	42.3
Albert Lea	5.42	*36.7*	4.77	39.9	—	—	7.22	41.5
St. Paul	5.58	*37.3*	4.49	38.6	—	—	6.31	38.3
St. Cloud	6.43	40.8	5.65	43.9	—	—	6.03	*37.2*
Mankato	5.42	*36.7*	4.78	39.9	—	—	7.03	40.9
Rochester	5.12	*35.4*	4.91	40.5	—	—	7.25	41.7
Bemidji	5.12	44.8	6.39	47.2	—	—	6.65	*39.5*
Duluth–Superior	4.12	31.3	—	—	$3.71	*30.4¢*	4.14	31.0

Ohio

From To	Brazil–Clinton (Indiana)		Ohio No. 8		Inner Crescent	
	Freight	Delv'd. cost	Freight	Delv'd. cost	Freight	Delv'd. cost
Oxford	$3.72	33.1¢	$4.28	32.5¢	$3.92	*29.6¢*
Harrison	3.56	32.4	4.28	32.5	4.26	*30.8*
Camden	3.72	33.1	4.28	32.5	4.21	*30.7*
New Paris	3.65	32.8	4.03	31.7	4.21	*30.7*

Indiana

From To	Brazil–Clinton (Indiana)		Ohio No. 8		Inner Crescent	
	Freight	Delv'd. cost	Freight	Delv'd. cost	Freight	Delv'd. cost
Butler	$3.59	32.6¢	$4.22	*32.4¢*	$4.72	32.5
Richmond	2.88	*29.5*	4.21	32.3	4.21	30.7
Ft. Wayne	3.28	*31.2*	4.22	32.4	4.72	32.5
Portland	3.18	*30.8*	4.22	32.4	4.72	32.5
Winchester	3.18	*30.8*	4.07	31.8	4.58	32.0
Union City	3.18	*30.8*	4.07	31.8	4.58	32.0
Decatur	3.13	*30.6*	4.22	32.4	4.72	32.5
N. Vernon	2.90	*29.8*	—	—	4.39	31.3
Madison	3.21	*30.0*	—	—	4.46	31.6

TABLE 44 *(continued)*

Tennessee

From / To	W. Kentucky Freight	W. Kentucky Delv'd. cost	Jellico, Tennessee Freight	Jellico, Tennessee Delv'd. cost
Union City	$3.72	29.7¢	$4.91	33.2¢
Dyersburg	3.72	29.7	4.91	33.2
Paris	3.22	27.7	4.50	31.7
Clarksville	2.34	24.0	3.91	29.6

Kentucky

From / To	W. Kentucky Freight	W. Kentucky Delv'd. cost	E. Kentucky Freight	E. Kentucky Delv'd. cost
Glasgow	$3.00	26.7¢	$3.49	28.5¢
Columbia	3.65	29.4	—	—
Lebanon	3.22	27.7	3.13	27.1
Springfield	2.96	26.6	3.07	26.9
Lexington	3.03	26.8	2.64	25.3
Shelbyville	3.14	27.3	3.02	26.7
Louisville	1.74	21.5	3.02	26.7

Missouri

From / To	S. Illinois Freight	S. Illinois Delv'd. cost	Belleville Freight	Belleville Delv'd. cost	Missouri Freight	Missouri Delv'd. cost
Hannibal	$2.98	28.2¢	$2.70	29.1¢	$2.04	27.5¢
Louisiana	3.74	31.3	3.35	31.9	2.75	30.7
St. Charles	3.19	29.1	2.86	29.8	3.42	33.7
Canton	3.18	29.0	2.90	30.0	2.12	27.8
Weldon	3.47	30.2	3.16	31.1	3.00	31.8

Iowa

From / To	S. Illinois Frt.	S. Illinois Delv'd. cost	Fulton–Peoria Frt.	Fulton–Peoria Delv'd. cost	Missouri Frt.	Missouri Delv'd. cost	Centerville Frt.	Centerville Delv'd. cost
Ft. Madison	—	—	$2.45	30.4¢	$3.17	32.6¢	—	—
Eddyville	$4.35	33.8¢	3.87	37.0	3.36	33.5	—	—
Dubuque	—	—	2.86	32.3	4.58	39.0	—	—
Cedar Falls	—	—	3.32	34.4	3.70	35.0	—	—

TABLE 44   (*continued*)

Iowa (*continued*)

To \ From	S. Illinois Frt.	S. Illinois Delv'd. cost	Fulton–Peoria Frt.	Fulton–Peoria Delv'd. cost	Missouri Frt.	Missouri Delv'd. cost	Centerville Frt.	Centerville Delv'd. cost
Ft. Dodge	$5.00	36.4¢	$4.26	38.8¢	$3.91	*35.9¢*	—	—
Sioux City	5.78	39.7	5.13	42.8	2.86	*31.2*	—	—
Council Blfs.	5.26	37.5	4.73	41.0	2.43	*29.2*	—	—
Creston	5.02	36.5	4.47	39.8	2.50	*29.6*	$2.36	33.4¢
Des Moines	4.67	35.1	4.12	38.2	3.28	*33.1*	—	—
Chariton	4.76	35.9	4.21	38.5	3.42	*33.7*	—	—
Marshalltown	4.67	35.1	3.82	36.7	3.68	*34.9*	—	—
Muscatine	—	—	2.00	*28.2*	3.88	35.8	—	—
Ottumwa	—	—	3.41	34.8	—	—	0.85	25.3

N.B. Italicized figures represent lowest delivered costs per million B.t.u. to each destination.

Rates used were those in effect June 1, 1960.

Sources: Midwest Coal Producers Institute, Inc., *Freight Schedules 2-A, 4, 5, 6, 7,* June 1, 1960; correspondence with Harold V. Scott, Traffic Manager, Midwest Coal Producers Institute, Inc.

Appendix B

In Chapter III it was estimated that the midwest's share of the national coal market will rise from 21.9 percent to 27.8 percent by 1979. But this estimate was predicated on several assumptions which require scrutiny.

The reliability of the assumption that sales to markets other than the utility market will remain constant is difficult to assess. Between 1939 and 1959 annual sales to these markets fell from 353 million to 246 million tons, most of the reduction stemming from losses in the retail and railroad fuel markets. Sales to these groups having reached a low level, the possibility of future tonnage losses has been minimized. In the absence of a major technological innovation, sales of coal for metallurgical purposes will grow steadily as the demand for steel expands, though at a rate slower than steel tonnage expansion because of continuing improvements in fuel efficiency. Markets abroad hold little promise of expansion and some danger of shrinkage as the competition from substitute fuels increases. Sales to manufacturing industries should remain fairly constant in the immediate future if collective efforts by the coal industry are successful in slowing down the movement away from coal to competitive fuels. The total effect of these divergent forces should be to maintain future sales to other-than-utility markets at about the current level.

The assertion that the midwestern coal industry's share of the national market will grow is also based on three additional assumptions: (1) that power production in the nation will continue to double every ten years as it has in the past several decades; (2) that the growth of power production in the midwest will keep pace with production in the other parts of the country where coal is used as an energy source; and (3) that the cost of producing power from atomic reactors will not be reduced sufficiently to diminish the demand for midwestern coal. If atomic power costs become competitive with the cost of power production in coal-fired plants, eastern coal fields fueling high-cost generating stations will be the first coal area adversely affected. This will have the effect of improving the midwest's relative position though it will pose a competitive threat with somber implications.

Selected Bibliography

Public Documents

Averitt, Paul, Louise R. Berryhill, and Dorothy A. Taylor. *Coal Resources of the United States.* U.S. Dept. of Interior Geological Survey. Geological Survey Circular no. 293, October 1, 1953.

Berquist, F. E., et al. *Economic Survey of the Bituminous Coal Industry Under Free Competition and Code Regulation.* National Recovery Administration, Division of Review, March, 1936.

Flynn, George J., Jr. *Average Heating Values of American Coals by Rank and by States.* U.S. Bureau of Mines Information Circular no. 7538, December, 1949.

Hotchkiss, Willard E., et al. *Mechanization, Employment, and Output per Man in Bituminous Coal Mining.* Vol. I. Works Projects Administration, National Research Project, 1939.

Illinois Department of Mines and Minerals. Annual Reports. Various years.

Kentucky Department of Mines and Minerals. Annual Reports. Various years.

National Resources Committee. *Energy Resources and National Policy.* Washington: U.S. Government Printing Office, January, 1939.

Office of Price Administration. *Preliminary Survey of Operating Data for Commercial Bituminous Coal Mines for Years 1943, 1944, 1945.* Economic Data Analysis Branch. O.P.A. Economic Data Series no. 2. Washington: U.S. Government Printing Office, n.d.

Office of Price Administration. *Survey of Commercial Bituminous Coal Mines, 1946.* Economic Data Analysis Branch. O.P.A. Economic Data Series no. 15. Washington: U.S. Government Printing Office, n.d.

Trapnell, W. C., and Ralph Ilsley. *The Bituminous Coal Industry with a Survey of Competing Fuels.* Federal Emergency Relief Administration. Washington, D.C., May 28, 1935.

U.S. Bureau of Census. *Census of Manufactures, 1954.* Vol. II.

U.S. Bureau of Mines. *Analyses of Tipple and Delivered Samples of Coal.* Bulletin no. 516. Washington: U.S. Government Printing Office, 1953.

―――― *Bituminous Coal and Lignite Distribution, 1959.* Mineral Market Survey Report no. 3035, March, 1960.

―――― *Bituminous Coal and Lignite in 1958.* Mineral Industry Surveys, Mineral Market Summary no. 2974, September 9, 1959.

―――― *Bituminous Coal Distribution by Market Areas, Calendar Year 1944, Year Ended September 30, 1945, and Coal Year 1945–46.* Mineral Market Survey no. 1500, March, 1947.

―――― *Mineral Facts and Problems.* Washington, D.C.: U.S. Government Printing Office, 1956.

―――― *Minerals Yearbook.* Various issues.

―――― *Preliminary Survey Producing, Administrative, and Selling Costs, 1936–42.*

―――― *Stocks and Consumption of Bituminous Coal and Lignite.* Mineral Industry Survey. Various issues, 1938–1959.

U.S. Coal Commission. *Report on the Effect of Irregular Operations on the Unit Cost of Production.* Washington: U.S. Government Printing Office, 1925.

U.S. Department of Interior. *Report to the Panel on Civilian Technology on Coal Slurry Pipelines,* May 1, 1962.

U.S. Federal Trade Commission. *Report on Corporate Mergers and Acquisitions.* Washington: U.S. Government Printing Office, 1955.

U.S. Federal Power Commission. *Statistics of Electric Utilities in the United States, 1961.*

―――― Steam Electric Plant Construction Costs and Annual Production Expenses. Annual Supplement, 1947–1959.

U.S. House of Representatives, Committee on Ways and Means. *Extension of Bituminous Coal Act of 1937.* 78th Congress, 1st Session, 1943.

―――― *Study of Monopoly Power.* Hearings before the Subcommittee on the Study of Monopoly Power of the Committee on the Judiciary. 81st Congress, 2nd Session, 1950.

U.S. Senate. *The Economics of Coal Traffic Flow.* Transportation Investigation and Research Board. Document no. 82. 79th Congress, 1st Session, 1945.

U.S. Temporary National Economic Committee. *Antitrust in Action.* Monograph no. 16. 76th Congress, 3rd Session, 1940.

―――― *Competition and Monopoly in American Industry.* Monograph no. 21. 76th Congress, 3rd Session, 1940.

―――― *The Structure of Industry.* Monograph no. 27. 76th Congress, 3rd Session, 1940.

Voskuil, Walter H. *The Competitive Position of Illinois Coal in the Illinois Coal Market Area.* Illinois State Geological Survey Bulletin no. 63. Urbana, Illinois, 1936.

Young, W. H., and R. L. Anderson. *Thickness of Bituminous Coal and Lignite Seams at All Mines and Thickness of Overburden at Strip Mines in the United States, in 1950.* Information Circular no. 7642.

Books

Bain, Joe S. *Barriers to New Competition.* Cambridge: Harvard University Press, 1956.

———— *The Economics of the Pacific Coast Petroleum Industry.* Part I: Market Structure. Berkeley: University of California Press, 1944.
———— *Industrial Organization.* New York: John Wiley & Sons, 1959.
———— *Price Theory.* New York: Henry Holt Co., 1952.
Baratz, Morton S. *The Union and the Coal Industry.* New Haven: Yale University Press, 1955.
Butters, J. Keith, John Lintner, and William L. Cary. *Effects of Taxation on Corporate Mergers.* Boston: Harvard Graduate School of Business Administration, Division of Research, 1951.
Christenson, C. L. *Economic Redevelopment in Bituminous Coal.* Cambridge: Harvard University Press, 1962.
Fisher, Waldo E., and Charles M. James. *Minimum Price Fixing in the Bituminous Coal Industry.* Princeton: Princeton University Press, 1955.
Graham, Herman D. *The Economics of Strip Coal Mining.* Bureau of Economic and Business Research. Bulletin no. 66. Urbana: University of Illinois, 1948.
Haynes, William W. *Present and Prospective Markets for West Kentucky Coal.* Bulletin no. 30, Bureau of Business Research. Lexington: University of Kentucky, August, 1955.
Healy, Kent T. *The Economics of Transportation in America.* New York: The Ronald Press, 1940.
Henderson, James M. *The Efficiency of the Coal Industry.* Cambridge: Harvard University Press, 1958.
Markham, Jesse William. *Competition in the Rayon Industry.* Cambridge: Harvard University Press, 1952.
Moore, Elwood S. *Coal.* 2nd Edition. New York: John Wiley & Sons, Inc., 1940.
National Bureau of Economic Research. *Business Concentration and Price Policy.* Princeton: Princeton University Press, 1955.
National Bureau of Economic Research. *Report of the Committee on Prices in the Bituminous Coal Industry.* Prepared for the Conference on Price Research. New York: National Bureau of Economic Research, 1938.
National Research Council. *Chemistry of Coal Utilization.* New York: John Wiley & Sons, Inc., 1945.
Nef, J. V. *The Rise of the British Coal Industry.* 2 Vols. London: G. Routledge & Sons, Ltd., 1932.
Nelson, Ralph L. *Merger Movements in American Industry, 1895–1956.* Princeton: Princeton University Press, 1959.
Parker, Glen L. *The Coal Industry.* Washington: American Council on Public Affairs, 1940.
Penrose, Edith T. *The Theory of the Growth of the Firm.* Oxford: Basil Blackwell, 1959.

Risser, Hubert E. *The Economics of the Coal Industry.* Lawrence, Kansas: Bureau of Business Research, University of Kansas, 1958.

Robinson, E. A. G. *The Structure of Competitive Industry.* London: Nisbet & Co., Ltd., 1935.

Schurr, Sam H. and Bruce C. Netschert. *Energy in the American Economy, 1850–1975.* Baltimore: The Johns Hopkins Press, 1960.

Shubin, John A. *Managerial and Industrial Economics.* New York: Ronald Press, 1961.

Sweezy, Paul M. *Monopoly and Competition in the English Coal Trade, 1550–1850.* Cambridge: Harvard University Press, 1938.

Weston, J. Fred. *The Role of Mergers in the Growth of Large Firms.* Berkeley: University of California Press, 1953.

Articles and Periodicals

Adelman, M. A. "Steel, Administered Prices and Inflation," *Quarterly Journal of Economics,* LXXV (February 1961), 16–40.

—— "The Measurement of Industrial Concentration," *Readings in Industrial Organization,* ed. George W. Stocking and Richard B. Heflebower. Homewood, Ill.: Richard D. Irwin, Inc., 1958.

Bain, Joe S. "Price and Production Policies," *A Survey of Contemporary Economics,* ed. Howard D. Ellis. Philadelphia: The Blakiston Company, 1949, pp. 129–173.

Campbell, Thomas C. *The Bituminous Coal Freight Rate Structure — An Economic Appraisal.* Bureau of Business Research. Morgantown, W. Va.: West Virginia University, June, 1954.

Christenson, C. Lawrence. "The Theory of the Offset Factor," *American Economic Review,* XLIII (September 1953), 513–547.

Fetter, Frank A. "The Economic Laws of Market Areas," *Quarterly Journal of Economics,* XXXVIII (May 1924), 520–530.

"A Half Century in Coal and the Next Ten Years," *Coal Age,* LXVI (October 1961).

Leonard, Norman H. "The Bituminous Coal Industry," *The Structure of American Industry,* ed. Walter Adams. New York: The Macmillan Co., 1950.

Mason, Edward S. "Price and Production Policies of Large Scale Enterprise," *American Economic Review,* Supplement, XXIX (March 1939).

Miller, John P. "The Pricing of Bituminous Coal. Some International Comparisons," *Public Policy,* ed. C. J. Friedrich, and Edward S. Mason. Cambridge: Harvard University Press, 1940, pp. 144–176.

Nerlove, Mark. "On the Efficiency of the Coal Industry," *Journal of Business* (July 1959).

Schmookler, Jacob. "The Bituminous Coal Industry," *The Structure of American Industry,* ed. Walter Adams, 1st ed. rev.; New York: The Macmillan Co., 1954.

Thompson, James H. *Markets and Marketing Methods of the West Virginia Coal Industry*. West Virginia University, Business and Economic Studies, II, no. 3. June, 1953.

Reports

Bituminous Coal Institute. *Comparative Fuel Costs, 1958*. Washington, D.C., n.d.
Coal Trade Association of Indiana. *Coal Production in Indiana, 1926–1950*, n.d.
———— *An Economic Survey of Coal*. Terre Haute, Ind.: Nov. 7, 1947.
Indiana Coal Association. *Report of Indiana Coal Production 1917–59*. Compiled annually. Terre Haute, Ind.
Keystone Coal Buyers Manual. New York: McGraw-Hill Publishing Co., 1958, 1959.
Keystone Coal Buyers Manual. *Coal Production in the United States*. New York: McGraw-Hill Publishing Co., June, 1960.
Moody's *Industrial Manual*, 1950–1961.
National Coal Association. *Bituminous Coal Data, 1959*. Washington: National Coal Association, 1960.
———— *Bituminous Coal Facts, 1960*.
———— *Steam–Electric Plant Factors, 1959*. Washington, D.C., July, 1960.
Standard and Poor. *Corporation Records*, 1961.
United Mine Workers of America. *Welfare and Retirement Fund: Four Year Summary and Chronology*. Washington, D.C., 1951.
Young, William Harvey. *Sources of Coal and Types of Stokers and Burners Used by Electric Public Utility Power Plants*. Washington: The Brookings Institution, Dec. 31, 1930.

Unpublished Material

Harline, Osmond LaVar. "Economics of the Indiana Coal Mining Industry." Unpub. diss., University of Indiana, 1958.
Tennessee Valley Authority. Bid sheets covering T.V.A. coal bids, furnished by Mr. E. C. Hill, Chief of Coal Procurement, T.V.A.

Index

Adelman, M. A., 64n, 162–63
Appalachian Coals, Incorporated, 151
Appalachian Wage Agreement, 110
Ayrshire Collieries, 65n, 80, 81, 84, 122, 123

Backward integration, 85–86; limiting factors, 86
Bain, Joe S., 3, 4, 5, 116n
Baratz, Morton, 142
Bargaining theory of price, 160
Barge, coal moved by, 16, 24
Barge rates, 24
Barriers to entry, 116–45; categories of, 116; effect of product differentiation on, 129–30
Bell and Zoller Coal Company, 65n, 77, 81
Binkley Coal Company, 76, 79, 81
Bituminous Coal Commission, 33
Bituminous Coal Institute, 49
Blackfoot Mining Company, 77
Bledsoe Coal Company, 77
Boiler equipment, effect on coal demand, 25
Bradbury Mine (Mid-West Utilities Coal Company), 77, 81
Bureau of Mines, 26
Burning equipment: influence on coal's desirability, 45–46; innovations in, 46; pulverized coal furnaces, 46; slag-type furnaces, 46; spreader stokers, 46; underfeed, 46
Butters, J. K., 72, 78
Buyer concentration, 87, 88; influence on prices, 87

Cady, G. H., 197
Campbell, T. C., 20
Captive coal, 36

Captive markets, 2
Captive mines, 85–86
Carmac Coal Company, 76
Cary, W. L., 72, 78
Central Competitive Field Interstate Agreement of 1898, 109
Central Illinois Electric and Gas Company, 89
Central Illinois Light Company, 89, 90
Central Illinois Public Service Company, 89, 131
Chamberlin, E. H., 4
Channels of distribution, effect on market power, 70
Chicago, Wilmington, and Franklin Coal Company, 77, 81
Christenson, C. L., 1, 108, 119n, 163, 201
Clark, J. M., 95–96
Cleveland Electric Illuminating Company, 137
Coal Age, 127
Coal-burning efficiency: in electricity production, 48; in iron-making, 48
Coal consumption, 27–37; domestic heating fuel market, 49; secular decline, 49
Coal demand: by coke plants, 43, 44; by electric utilities, 44; elasticities of, 3, 53–59, 59–61; export market, 43, 44; metallurgical, 56; midwestern, stability in electric utility market, 44; national market, 45; projected for midwestern industry, 213; secular, 41, 44; shift in, 41
Coal distribution, by states, 27–37. *See also* Coal consumption
Coal pipelines, 137, 140
Coal preparation, 91, 93; changes in output mechanically cleaned, 111